# VOICES OF THE KNOX MINE DISASTER

STORIES, REMEMBRANCES, AND REFLECTIONS OF THE
ANTHRACITE COAL INDUSTRY'S LAST MAJOR
CATASTROPHE, JANUARY 22, 1959

*Robert P. Wolensky*

Robert P. Wolensky
University of Wisconsin-Stevens Point

Nicole H. Wolensky
University of Iowa

Kenneth C. Wolensky
Pennsylvania Historical and Museum Commission

Commonwealth of Pennsylvania
Pennsylvania Historical and Museum Commission
2005

## COMMONWEALTH OF PENNSYLVANIA

Edward G. Rendell, Governor

## PENNSYLVANIA HISTORICAL AND MUSEUM COMMISSION

Wayne S. Spilove, Chairman

Rhonda R. Cohen

Lawrence A. Curry, Representative

Jane M. Earll, Senator

Gordon A. Haaland

Robert A. Janosov

Janet S. Klein

Stephen R. Maitland, Representative

Cheryl McClenney-Brooker

Brian C. Mitchell

Kathleen A. Pavelko

Allyson Y. Schwartz, Senator

Francis V. Barnes
Secretary of Education
ex officio

Barbara Franco, Executive Director

Copyright © 2005 Commonwealth of Pennsylvania
ISBN 0-89271-114-0

# **D**EDICATION

To the families and friends of the twelve victims and to the citizens and institutions of Pennsylvania who have kept memory of the Knox Mine Disaster, as well as the heritage of the anthracite era, alive.

*The principal factor that threatens to cut short the life of the anthracite industry, to curtail production, and to affect the economic structure of the people and businesses dependent on anthracite for their livelihood, is inundation of anthracite mines.*
Solomon H. Ash, "Buried Valley of the Susquehanna," U.S. Bureau of Mines, *Bulletin* 494, 1950.


*The world notices some people and it doesn't know much about others. That's a truism, but it's something that historians should worry about, because if they don't watch out, they're going to overlook a lot that happens.*
Dorothy Day quoted by Robert Coles, 1998

*History, despite its wrenching pain,*
*Cannot be unlived, and if faced*
*With courage, need not be lived again.*
*Lift up your eyes upon*
*This day breaking for you.*
*Give birth again*
*To the dream.*
Maya Angelou, January 20, 1993


*It is possible for people to make something of themselves other than that which history has made of them.*
E.P. Thompson

# Table of Contents

**Acknowledgments** .................................................. vii

**Introduction** ........................................................ ix

**Chapter 1.** The Knox Mine Disaster,
January 22, 1959 ................................................ 1

**Chapter 2.** Retreats and Rescues ................................ 21

**Chapter 3.** Delayed Escapes and Missing Men ............. 63

**Chapter 4.** Spouses and Siblings ................................ 101

**Chapter 5.** The Children ........................................... 145

**Chapter 6.** Remembering the Knox Mine Disaster:
Poets, Composers, and Writers ............................ 195

**Chapter 7.** Memorializing and Reflecting on the
Knox Mine Disaster and Anthracite History ........ 227

**Glossary of Anthracite Mining Terms** ..................... 257

**Index** .................................................................... 263

**About the Authors** ................................................ 268

# ACKNOWLEDGMENTS

We would like to express our sincere gratitude to several persons who helped with research, writing, and editing: Jack Larsen, Ann Larsen, Bill Hastie, Ron Slusser, Matt Slusser, Judy Kelly, Jim Kelly, Chad Balke, Megan Heurion, Lars Higdon, and Nicole Eckes. We are indebted to Steve Lukasik, George Harvan, John Dziak, and William Best for granting permission to use several photographs, and to songwriters Adrian Mark Bianconi, Frank Murman, Bob Rogers, Lex Romaine, and Ray Stephens for permission to include lyrics from their compositions. Five area newspapers provided access to the Knox-related stories, poems, letters, and photographs: the *Citizens' Voice (Wilkes-Barre)*, the *Scranton Times*, the *Times Leader* (Wilkes-Barre), the *Sunday Independent* (Wilkes-Barre; no longer published), and the *Sunday Dispatch* (Pittston).

We would also like to thank the following organizations for providing various forms of support: the Center for the Small City at the University of Wisconsin-Stevens Point, the Department of Sociology at the University of Wisconsin-Stevens Point, the College of Letters and Science the University of Wisconsin-Stevens Point, the Center for Twenty-first Century Studies at the University of Wisconsin-Milwaukee, the Pennsylvania Historical and Museum Commission, Wilkes University, the Luzerne County Historical Society, the Anthracite Heritage Museum, the Greater Pittston Historical Society, the Osterhaut Memorial Library, and the Institute for Research in the Humanities at the University of Wisconsin-Madison.

We wish to convey our heartfelt appreciation to the members of the Altieri, Baloga, Boyar, Burns, Ostrowski, Sinclair, Featherman, Gizenski, Orlowski, and Stefanides families who graciously contributed oral histories, personal letters, newspaper clippings, and other documents related to the deaths of their loved ones in the Knox Mine Disaster; and to the surviving mineworkers and other persons who provided the oral memoirs that allowed us to more fully appreciate the human tragedy that unfolded at the River Slope Mine on January 22, 1959.

Thanks to Diane B. Reed, chief of publications, and Susan Gahres, who designed the book, of the Pennsylvania Historical and Museum Commission's publications program. We owe them our gratitude for the diligence and professionalism they have displayed in working on the book.

Finally, we want to express sincere thanks to our families and friends for their support and encouragement. We truly couldn't have undertaken and completed this project without you.

# Introduction

    This volume is intended to accompany our earlier book on the Knox Mine Disaster published in 1999 by the Pennsylvania Historical and Museum Commission. The response to that historical-analytical account, now in a third printing, coupled with the voluminous research materials in our possession, prompted us to consider presenting the story of anthracite's last major catastrophe from another viewpoint—one that is more humanistic and story-based.

    We wanted the mineworkers to relate the saga of their escape and the widows and other family members to convey the sorrow of their loss. We wanted the reader to relive the emergency and the aftermath through the words of those who lived it. We wanted to highlight the dedication of the local community's remembrance of the epic event.

    In sum, our purpose was to let the voices of the Knox Mine Disaster speak. The volume consists of first-person accounts (oral and written) of the tragedy, its consequences, and meaning (chapters one through five), and of the various ways in which the community has remembered the calamity over the past forty-five years (chapters six and seven). We drew mainly on oral history interviews conducted with survivors, family members, mineworkers, and other knowledgeable persons. These audiotape-recorded memoirs are part of The Northeastern Pennsylvania Oral and Life History Project.[1]

Oral history has gained wide acceptance in recent years as a research method in its own right and as a supplement to documentary, survey, and other types of evidence. Notwithstanding the problems associated with human memory, oral history has been particularly effective in gathering first-hand information when other sources fall short. For example, the four government investigations following the disaster heard testimony from dozens of mineworkers and company supervisors regarding the escape of sixty-nine men from the mine, but none of the legally grounded interrogations could convey the depth of feeling captured in the narrative excerpts presented here in chapter two and chapter three. Similarly, spouses and children were called to testify before the official inquiries, but the legalistic nature of the dialogue did not permit the level of emotion expressed by those who contributed oral histories in chapter four and chapter five.[2]

A word must be said about the methods used in excerpting and editing the interviews. To facilitate readability we eliminated many of the questions asked by the interviewer and corrected grammar in most cases, although colloquial words and phrases were usually left in. We also rearranged some sentences or paragraphs to maintain the logical flow of commentary; ellipses indicate where this has been done. Ellipses are also used, as usual, to indicate where we have cut the dialogue. In making such changes we made every effort to remain faithful to the meanings intended by the narrator. When a person cited an incorrect number or statement we included the correction in brackets. Finally, we left untouched the colloquial and technical mining terms but provided brief definitions in parentheses and fuller explanations in the glossary.

Another word must be said about the items included. We selected excerpts and writings that addressed three questions. How did mineworkers survive the disaster? What did it mean to wives, children, brothers, sisters, and fellow workers? How have poets, lyricists, essayists, historians, songwriters, and eulogists remembered it? In all cases, we were looking for voices that spoke to the human responses to the loss and trauma.

Although we relied mainly on oral histories to convey the human consequences of the disaster, the "voices" of newspaper reporters, editorialists, and letters to the editor writers were included to supplement the spoken accounts. These writers, we believe, help take us back to the period so we can more fully grasp the disaster's personal and familial repercussions. Yet, because of space limitations we were forced to leave out most of the written accounts we had collected. For example, we eliminated a chapter that presented a sample of the hundreds of news stories, editorials, and letters to the editor that captured the incredibly angry public reaction to the calamity and its aftermath.

Furthermore, we had to exclude numerous oral histories. For example, we did not incorporate the interviews with the workers who plugged the hole in the Susquehanna River; or the attorneys and judges who prosecuted, defended, and presided at the Knox-related trials; or the many mineworkers employed by Knox and other companies. These voices will have to wait for another time and forum.

Despite the omissions, we hope that the voices we have included will complement and expand upon our earlier effort to historically document and critically analyze anthracite coal's last major cataclysm.

## NOTES FOR THE INTRODUCTION

1. A full list of the persons interviewed for our Knox Mine Disaster research can be found in appendix I of *The Knox Mine Disaster: January 22, 1959*. Dr. Ellis Roberts conducted one of the interviews, with mineworkers Joseph Kaloge, excerpted in chapter two.

2. For more on oral history see Michael Frisch, *A Shared Authority* (Albany: State University of New York Press, 1990); Ronald J. Grele, Envelopes of Sound: *The Art of Oral History: Essays on the Craft and Meaning of Oral and Public History* (Westport, Conn.: Greenwood Press, 1991); Alessandro. Portelli, *The Death of Luigi Trastulli and Other Stories: Form and Meaning in Oral History* (Albany: State University of New York Press, 1991); *Alessandro Portelli, The Battle of Valle Giulia: Oral History and the Art of Dialogue* (Madison: University of Wisconsin Press, 1997); Robert Perks and Alistair Thompson, *The Oral History Reader* (New York: Routledge, 1998); Paul Thompson, *The Voice of the Past: Oral History* (New York: Oxford University Press, 2000); Donald A. Ritchie, *Doing Oral History: A Practical Guide* (New York: Oxford University Press, 2003).

# Chapter One
## The Knox Mine Disaster
## January 22, 1959[1]

*They didn't think the river would cave in. They were living in a fool's paradise, even making plans to do more mining in that area.*
William Hastie, laborer, Knox Coal Company

*One of these days we are all going to die with our boots on like rats in a trap.*
Frank Orlowski, victim, Knox mine disaster

In late January 1959, after two days of steady rain and sixty-degree temperatures, the Susquehanna River began to thaw. Almost overnight it turned into a surging, iceberg-laden torrent as it traveled through northeastern Pennsylvania. Observers at the recording station in the Wyoming Valley city of Wilkes-Barre, in Luzerne County, watched as the river rose from 2.1 feet on January 20, to just below the twenty-two foot flood stage on the night of January 23.

The rising current would prove catastrophic for the Knox Coal

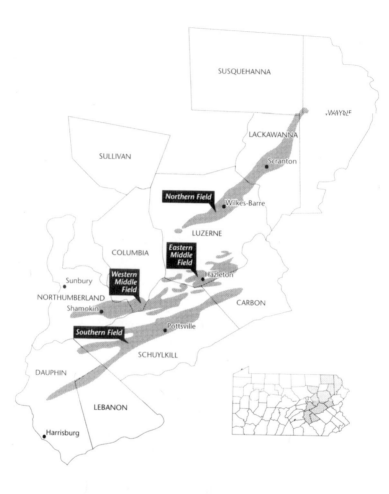

Fig. 1. The Anthracite Coal Region of northeastern Pennsylvania. (From E Willard Miller, *A Geography of Pennsylvania*, Penn State University Press, 1995)

Company's operations in Port Griffith, a small town seven miles upriver from Wilkes-Barre and ten miles south of Scranton in Lackawanna County in the northern anthracite coal field (figure 1). The company's River Slope Mine lay near the eastern shore of the Susquehanna River, and many of its workings were under the river. The mine employed 174 men, many of whom had voiced fears about their workplaces. Laborer Joe Poluske complained about the raincoats and boots he had to wear because "droppers" (large droplets of water) constantly fell from the roof.[2] Fear of drowning dogged electrician Herman Zelonis who told his sister, "If that river comes in, we'll be drowned like rats." Another laborer, Frank Orlowski, forewarned his sister that, "one of these days we are all going to die with our boots on like rats in a trap." And rockman Gene Ostrowski had nightmares about his bedroom ceiling cracking and falling on him and his young son as they slept together in the same bed. Orlowski, Zelonis, and Ostrowski were among the Knox Mine Disaster's victims when the ceiling in the River Slope gave way and the river crashed through.[3]

## Disaster at the River Slope

On January 22, 1959 at 7:00 a.m., eighty-one men reported for the first of three shifts. Seventy-five were employed by the Knox Coal Company, three by the Stuart Creasing Company which had subcontracted with Knox to dig an internal rock tunnel, and three by the Pennsylvania Coal Company, lessor of the mine to Knox. Seventy-two of the Knox workers headed to the current center of mining activity in the May Shaft. Two of the Pennsylvania men traveled to the bottom-most vein, the Red Ash, to repair a pump, while the third began surveying the May Shaft section.

The six remaining men traveled to the River Slope section. Three of them were Knox employees who were removing equipment from the top-lying Pittston Vein, or the Big Vein as the miners called it, whose coal had recently been exhausted. The other three were rockmen from the Creasing firm who were in the last few hours of completing a tunnel connecting two veins—the Pittston and the Marcy.

Assistant foreman Jack Williams, a sixty-two-year-old Scottish-born mining veteran, and laborers Fred Bohn and Frank Domoracki were the Knox workers at the River Slope, while Steward Creasing's rockmen included Eugene Ostrowski, Charles Featherman, and Joseph "Tiny" Gizenski. At about 11:30 a.m., Bohn and Domoracki summoned Williams from his conversation with the rockmen so he could examine the sharp cracking sounds they heard coming from the wooden roof supports or props in a particular chamber. "I no more than put my foot in the place and looked up," said Williams in a characteristic burr before a state investigating committee, "than the roof gave way. It sounded like thunder. Water poured down like Niagara Falls."[4]

The three awestruck workers sprinted some two hundred feet up the slope to save their lives. Once outside, they hurriedly phoned superintendent Robert Groves on a company line and relayed the distressing news. Groves immediately called the underground stations and ordered a full evacuation. To avoid panic, Groves reportedly withheld the cause of the emergency. The three rockmen were not as fortunate as their Knox colleagues. The nearly completed tunnel was located directly below the breach. It quickly and forcefully filled with water, and the rockmen drowned.

### NARROW ESCAPES AND TRAGIC DISSOLUTION

After receiving Groves's order, thirty-three men made their way to elevators at the May and Hoyt Shafts. Forty-five others were unable to reach the lifts and were trapped deep in the pit. All would surely drown unless they could find an exit or unless mining officials could speedily cap the gaping hole that was whirling and growing larger along the river's eastern bank (figures 2 and 3).

At about 2:45 p.m., as government authorities huddled with local mining experts to plot an emergency strategy, a sodden laborer named Amadeo Pancotti made a perilous ascent out of the Eagle Air Shaft. Situated a few hundred feet upriver from the breach, this abandoned air tunnel became the only exit available. Thirty-two additional workers surfaced through this outlet in two separate groups. Pancotti traveled with the first crew consisting of seven men led by Pacifico "Joe" Stella, a thirty-five-year-old surveyor employed by the Pennsylvania Coal Company. Stella not only knew the mines from his many inspections, but possessed maps that allowed his group to chart a direct course to the opening.

Myron Thomas, a Knox assistant foreman, led the other group of twenty-five men. They became lost when they separated from Stella's con-

Fig. 2. The whirlpool in the Susquehanna River at the site of the Knox Mine Disaster in Port Griffith, Pennsylvania. (Courtesy of *Citizens' Voice*)

tingent in a parting, the nature of which involves some controversy. Without adequate maps Thomas' group wandered for nearly seven hours until discovered by a search party. Chapters two and three present first-person oral history accounts from members of the Stella and Thomas parties.

## THE TWELVE VICTIMS OF THE KNOX MINE DISASTER

The survival of thirty-three men at the Eagle Air Shaft meant that twelve were still missing. Family members gathered at the site with hundreds of others to watch and hope for a miracle. Veteran mineworkers theorized that some of the trapped could survive for a short time if they could find an air pocket. But how long would they last without food or water or, eventually, air? Hope nevertheless remained alive. Fr. Edmund Langan of St. John's Church in Pittston arrived to bless the mine and lead prayers. The next day, January 23, water continued to flow underground, and methane gas began rising from the mine. Government officials decided to halt all rescue operations. There would be no miracle.

Some of the deceased had delayed leaving, no doubt because they did

Fig. 3. Newspaper account of the Knox Mine Disaster. (Courtesy of Pittston *Sunday Dispatch*)

| Victim | Age | Place of Residence | Occupation | Number In Family[a] | Years Experience |
|---|---|---|---|---|---|
| Samuel Altieri[b] | 62 | Hughestown | Electrician | 7 | 16 |
| John Baloga[b] | 54 | Port Griffith | Miner | 5 | 35 |
| Benjamin Boyar[c] | 55 | Forty Fort | Electrical Foreman | 3 | 28 |
| Francis Burns[c] | 62 | Pittston | Lessor's Inspector | 4 | 40 |
| Charles Featherman[d] | 37 | Muhlenburg | Rockman/ Laborer | 2 | 12 |
| Joseph Gizenski[d] | 37 | Hunlock Creek | Rockman/Foreman | 4 | 22 |
| Dominick Kaveliskie[b] | 52 | Pittston | Laborer | 2 | 25 |
| Frank Orlowski[b] | 42 | Dupont | Laborer | 3 | 20 |
| Eugene Ostrowski[d] | 34 | Wanamie | Rockman/Laborer | 4 | 16 |
| William Sinclair[b] | 48 | Pittston | Laborer | 2 | 35 |
| Daniel Stefanides[b] | 33 | Swoyersville | Laborer | 5 | 2 |
| Herman Zelonis[b] | 58 | Pittston | Electrician | 1 | 40 |

[a]Immediate family, excluding the deceased; not all were dependents
[b]Employee of the Knox Coal Company
[c]Employee of the Pennsylvania Coal Company
[d]Employee of the Stuart Creasing Company, Rock Contractors

Fig. 4. The twelve victims of the Knox Mine Disaster.

not know the nature or extent of the emergency. Motorman Sam Altieri took superintendent Groves' call and traveled further into the pit to notify others. Herman Zelonis decided to change clothes. John Baloga put his tools away, probably expecting to return to his workplace in a day or two. William Sinclair climbed up a chamber to warn fellow crew members, but by the time he returned and started toward the May Shaft, the water had risen such that neither he nor his buddy, Daniel Stefanides, could reach safety. Their crew chief, miner Michael Lucas, did escape, and tells his story in chapter one. In chapters four and five, Sinclair and Stefanides family members reflect on the loss of their loved ones.

Five men had no real chance of survival. The three rockmen—Ostrowski, Featherman, and Gizenski—were certainly the first to drown as the explosion of water from the river flooded their workplace. The two others were Pennsylvania Coal Company workers, Francis Burns and Benjamin Boyar, who were trapped in the Red Ash Vein. Phone connections had not been maintained in the deepest reaches of the mine and, even if they had been, as William Hastie suggests in chapter one, Burns and Boyar had little chance of egress because of the flooding in the narrow man-ways leading to their work level. Oral history recollections by family members of these five victims are presented in chapters four and five.

The deceased workers ranged in age from thirty-three to sixty-two. Their experience underground varied from two to forty years. All but one was married (figure 4). Their ethnic diversity—Irish, Italian, Lithuanian, Polish, Scottish, and Slovak—reflects the rich ethnic heritage of the anthracite region.

## Plugging the Hole and Sealing the Breach

During the peak of the emergency, an estimated 2.7 million gallons of water per minute streamed underground. U.S. Geological Survey instruments indicated that the water pool in the River Slope reached 502 feet above sea level at 8 p.m. on January 25, a figure significantly higher than the usual height of 100 feet. In total, 10.37 billion gallons coursed into the River Slope and surrounding mines.

Work crews toiled around the clock for three days to plug the cavity. They cut and bent the railroad tracks toward the river and used a railroad engine to push about sixty coal hopper cars—fifty-ton behemoths called

Fig. 5. Plugging the hole in the Susquehanna River at the site of the Knox Mine Disaster. (Courtesy of George Harvan)

Fig. 6. Securing the breached area after the hole had been filled. (Courtesy of Stephen Lukasik)

gondolas—into the void. With the help of cranes and large dump trucks they added about four hundred one-ton coal cars as well as some 25,000 cubic yards of dirt, rock, and boulders. Finally, the whirlpool stopped. Yet the temporary patch could not stop the water from seeping underground. As late as March 23, the flow amounted to 20,000 gallons per minute (figures 5 and 6).

The Pennsylvania Department of Mines and Mineral Industries began a massive pumping operation to save the adjoining mines and, ultimately, the Wyoming Valley anthracite industry, since many workings were interconnected. The state provided funding to place forty siphons in strategic mine shafts to drain billions of gallons from the earth. After a few weeks of pumping, a search team ventured underground to inspect the temporary patch and look for bodies. None were found.

In early spring 1959, construction crews began work on a permanent seal. They diverted the river and built an earthen cofferdam around the hole. They drained the dam to expose the river bottom and then drilled several boreholes into the mine through which they poured 1,200 cubic

Fig. 7. Building a cofferdam to seal the River Slope Mine. (Courtesy of William Hastie)

yards of concrete and 26,000 cubic yards of sand. The federal and state governments allocated nearly $5 million for the project. The Commonwealth of Pennsylvania presented bills totaling $1.5 million to the Knox and the Pennsylvania Coal Companies, but Knox paid nothing because it declared bankruptcy, and the Pennsylvania successfully argued that, as the lessor, it was not liable (figure 7).

As the water flowed out of the River Slope into adjoining workings, mine after mine closed down. The Pennsylvania Coal Company and the Lehigh Valley Coal Company owned most of the mines in the area and decided to permanently close all operations, most of which had been leased out. Within a matter of months, all mines in the middle portion of the northern field, around the city of Pittston, were permanently idled (figure 8)

## The Causes of the Knox Disaster

Investigations by federal, state, and county authorities found the disaster's immediate cause in two illegal chambers and three connecting crosscuts dug under the river well past officially designated "stop lines." The chambers were quarried without the required benefit of surface boreholes to determine the thickness of the rock cover (thirty-five feet was the minimum) and without proper surveying. Mining under the supervision of company bosses, Knox mineworkers followed the thick Pittston Vein at a sharp upward angle toward the riverbed until the rock cover dwindled to somewhere between nineteen inches and a few feet. The thin roof could not withstand the added weight of the surging Susquehanna River.

Federal and state mine inspectors had discovered the illegal mining months earlier, as had surveyors from the Pennsylvania Coal Company. They informed officials at both the Knox and the Pennsylvania Coal Companies but took no official action until mid-January 1959. On January 13 the federal government inspector finally ordered all work stopped. Yet on that date, Knox supervisors allowed the night shift to take thirteen additional cars of coal. The rupture occurred at the precise spot of the last shift's diggings.

Fig. 8. Mining operations in the middle portion of the Northern Anthracite Field are idled. (Courtesy of *Sunday Independent*)

Fig. 9. Assistant foreman William Receski testifies before an investigative body of the Pennsylvania legislature. (Courtesy of Stephen Lukasik)

Why would company officials and miners with years of experience take such risks? Greed stands as one answer. The Big Vein glistened at twelve to fifteen feet of pure coal. Both the company and the workers profited from the high output. Weak mining laws and enforcement represented another explanation. State law at the time mandated only a $500 fine and a ninety-day prison term for illegal mining.

## Corruption in the Anthracite Industry

Two additional factors must also be considered. Corruption within the anthracite industry had become widespread by 1959. When allegations surfaced regarding organized crime's influence at the Knox Coal Company and in District 1 of the United Mine Workers of America (UMWA), the U.S. attorney general's "Special Group on Organized Crime" convened a grand jury. After hearing from over thirty witnesses and reviewing voluminous records from the Knox and Pennsylvania Coal Companies, UMWA District 1, and three local banks, the grand jury handed down indictments against several individuals.

Among them were two owners of the Knox Coal Company. Robert Dougherty, president and general manager, was indicted on twenty-five counts, and Louis Fabrizio, secretary-treasurer, received nine counts. They were charged with bribing UMWA Local 8005 officials in a "sweetheart" deal, whereby the union disregarded several aspects of the labor-management contract.

The grand jury delivered a thirty-four count indictment against Dominick Alaimo, an alleged member of organized crime and a committeeman at Local 8005, for receiving the illegal payments from Knox officials. Thirty-four counts were also directed at Charles Piasecki, president of the local, and at Anthony Argo, another committeeman. Local 8005 had jurisdiction over ten collieries in Luzerne and Lackawanna Counties.

In August 1959, the Grand Jury indicted fifty-eight-year-old August J. Lippi, president of UMWA District 1, for accepting illegal payments from Knox. Astonishingly, the investigation also revealed that Lippi had been a part owner of the Knox Coal Company in blatant violation of the Taft-Hartley labor law. John Sciandra, whom the Pennsylvania Crime Commission called a former head of the region's organized crime family, had been a colleague of Lippi and an original part owner of Knox, although he had died in 1949 and passed his shares to his wife, Josephine. Although she was also an owner, she escaped indictment in this phase of the inquiries.

Three other bodies investigated the disaster. The U.S. Bureau of Mines participated in hearings convened by a specially appointed Commission of Mine Inspectors from the Pennsylvania Department of Mines and Mineral Industries. Twenty-two witnesses testified in February and March 1959. Each body produced a detailed report, with the state recommending criminal charges against specific persons. A special committee of legislators, appointed by the Pennsylvania General Assembly, carried out another inquisition (figure 9). After listening to sixty-four witnesses, the Legislative Committee to Investigate the Knox Mine Disaster issued a final report on July 27, 1959. Disappointingly, it held no one responsible but instead blamed, "a chain of events each link of which was necessary or incidental to the resultant tragedy."[5]

The Commonwealth of Pennsylvania, under the leadership of Luzerne County District Attorney Albert Aston and his successor, Stephen Teller, indicted and prosecuted Lippi, Fabrizio, and Doughery, as well as superintendent Robert Groves, assistant foreman William Receski, and Pennsylvania

Fig. 10. August Lippi going to trial for income tax evasion in 1963. (Courtesy of *Scranton Times*)

Coal Company employees Fritz Renner and Ralph Fries. The charges ranged from mining and labor law violations to conspiracy and manslaughter.

After a series of state and federal trials in Wilkes-Barre, Scranton, Easton, and Wilmington, Delaware, the legal proceedings yielded few penalties. Union officials Alaimo and Piasecki were convicted of accepting bribes and given two-year prison sentences plus a fine, while Argo pleaded guilty to a reduced fourteen-count charge and received a fine plus a suspended sentence. Lippi, Dougherty, and Fabrizio, who presumably paid the bribes, had their convictions overturned on appeal. Groves and Receski were found not guilty in their cases, as were Fries and Renner of the Pennsylvania Coal Company. In the final analysis, no persons or companies were held responsible for the illegal mining or the death of twelve victims.[6]

However, the four owners were later convicted in federal court for income tax evasion. Fabrizio received a prison sentence of 181 days; Dougherty was sentenced for one year plus a $10,000 fine (he served only four and one-half months); Sciandra was fined $2,000 and placed on probation; and Lippi took a $5,000 fine and a sentence of three years imprisonment plus three years probation. Lippi was also convicted of bank fraud in a related case, which brought a five years sentence plus a $10,000 fine. His sentences ran concurrently and he served only forty-one months. Because of a bankruptcy declaration, Lippi never paid the fine or, as it turned out, an income tax bill totaling $337,334 (figure 10).

The second underlying cause of the disaster related to the reorganization of the anthracite industry during the first half of the twentieth century. In the late 1890s, two of the large operations, the Pennsylvania Coal Company and the Hudson Coal Company, began subcontracting mine sections to individual miners and other mining entrepreneurs. The subcontractors hired small crews and found ways to cut costs, especially labor costs, by forcing greater productivity from workers. Some ignored safety rules, the labor-management agreement, and regulations against mining in off-limit places. Subcontracting expanded during the first two decades of the century and precipitated two major strikes at the Pennsylvania Coal Company, in 1920 and 1928. In the 1930s, the Pennsylvania Coal Company went beyond subcontracting and began leasing large sections and even entire collieries to newly formed independent coal companies. The tenants were required to sell the coal to the lessor who then processed it, shipped it over an affiliated railroad, and sold it at market. Two of the area's other large companies, the Lehigh Valley Coal Company and the Hudson Coal Company, soon followed suit.

Both forms of tenancy—the subcontract and the lease—provided significant benefits to the large firms. They lowered production costs, as subcontractors and leaseholders were forced to bid against one another, often at the expense of workers' pay and mine safety. They boosted output as work and safety rules were cast aside and production incentives were offered to workers. Moreover, the arrangement allowed the lessor to retain control over key aspects of the coal business while avoiding the actual mining, thereby lessening the possibility of major conflict with the historically militant anthracite work force. Finally, over the course of time, tenancy facilitated the neutralization and eventually the corruption of UMWA District 1.[7]

By the late 1930s the subcontracting and lease-holding system had inaugurated a new era in northern field coal mining. Safety became less important than output, and widespread corruption launched a regime of illegal mining, bogus inspections, kickbacks for subcontracts and leases, and other forms of systematic criminal activity. As the leader in issuing tenancy agreements, the Pennsylvania Coal Company had sublet virtually all of its mines and collieries by the early 1940s. Alleged members of organized crime were prominent among the company's lessees—including some of the principals at the Knox Coal Company.

The reorganization from the large corporations to the smaller subcontractors and leaseholders exacerbated the field's already tense labor-management relations. The decline in demand for anthracite following the major strike in 1925–26 had increased unemployment, and the drive by the companies to cut costs and change the work culture through devices like tenancy led to a major labor rebellion. As workers looked to the UMWA for support and found little, they formed an alternative union in 1932 called the United Anthracite Miners of Pennsylvania (UAMP). Under the leadership of Rinaldo Cappellini, District 1 president in the 1920s, and Thomas Maloney, a UMWA local president from Wilkes-Barre, the upstart union garnered the support of well-known labor priest Rev. John J. Curran of Wilkes-Barre, who had helped President Theodore Roosevelt settle the 1902 strike.

The rebels brought the industry to a virtual standstill with a series of strikes between 1931 and 1934. Attempts at negotiations involving the national administration and powerful senator Robert Wagner of New York failed to end the strife. But by 1935, the combined forces of the companies, the UMWA, and the Roosevelt Administration forced the UAMP's capitulation. With organized labor "tamed," the subcontracting-leasing system expanded. In one sense, therefore, the Knox mine disaster can be viewed as an outcome of the anthracite industry's new mode of operation.[8]

## THE AFTERMATH OF THE KNOX MINE DISASTER

Despite a decline that had begun in the mid-1920s, the hard coal business still commanded a large share of the local economy in 1959. One estimate of the disaster's impact put the direct and indirect job loss at 7,500 and the payroll withdrawal at $32 million. In spite of losses, deep mining continued into the early 1970s in the upper reaches of the northern field above Scranton (whose mines were unaffected by the Knox disaster) and in the southern areas below Wilkes-Barre. By the 1970s, however, the high cost of mine pumping in the Wyoming Valley, along with anthracite's declining competitive position and growing corruption, brought the end of the deep mining era. Billions of tons of anthracite remain in the ground.

In Luzerne County, mining employment plunged from 10,200 in 1958 to 2,100 in 1970. During the same period Lackawanna County saw a decrease from 4,000 to 500. The final phase of anthracite deindustrialization seemed unstoppable. The anthracite region became known as a "depressed area." To address the crisis, leaders from government, labor, and business joined forces to remake the local economy. Building on earlier programs that recognized the area's over-dependence on coal, leaders in Scranton, Wilkes-Barre, Hazleton, Pottsville, and other cities initiated major development initiatives. The drives attracted numerous manufacturing and service businesses.[9] Despite the economic bolstering, the out-migration of large numbers continued as residents moved to other states and regions to secure employment.

The Knox disaster resulted in changes to Pennsylvania's mining laws. Legislators introduced a series of bills to strengthen mining regulations. Among the most important was the prohibition of mining under rivers, streams, and waterways without prior approval of Department of Mines and Mineral Industries.

Another consequence of the disaster has been apparent in the social and cultural realms. Its enduring effect on the community can be seen in regular commemorations and news stories, community forums and lectures, historical markers and monuments, books and articles, and television documentaries and radio interviews. Indeed, the interest seems to grow with each passing year. These remembrances confirm the tragedy's enduring hold on the citizenry. The poems, songs, oral histories, and essays in chapter six and in chapter seven address this important dimension of the Knox legacy.

Another important aspect can be seen in the broader social and cultural meanings the community has brought to the disaster and, indeed, the anthracite industry. As noted in chapters six and seven, for many people Knox has come to represent the exploitative mining and employment practices of the entire hard coal business. It, therefore, stands as a symbol of a dangerous occupation that took the lives of thirty-five thousand men and boys. It exemplifies the often dishonest mining companies who made billions of dollars yet begrudged mineworkers living wages and safe conditions. It signifies a feudal economic past, "a state called anthracite," where the powerful owners and their agents ruled over the weaker workers and their families in all aspects of life—employment, housing, politics, consumption, and even death.[10]

Thus the Knox Mine Disaster has remained alive in the hearts and minds of so many people because it speaks to deeply held beliefs regarding personal identity and regional heritage. As such, it remains much too important to forget. Indeed, it stands as an event that deserves foremost recognition and remembrance.

### Notes for Chapter One

1. For full details on the causes and consequences of the disaster see Robert P. Wolensky, Kenneth C. Wolensky, Nicole H. Wolensky, *The Knox Mine Disaster: January 22, 1959—The Final Years of the Northern Anthracite Industry and the Effort to Rebuild a Regional Economy* (Harrisburg: Pennsylvania Historical and Museum Commission, 1999).

2. The appendix contains a glossary of mining terms such as "droppers" and "roof." Brief definitions of some mining terms have been included in parentheses in the text.

3. Joseph Poluske, Testimony, Joint Legislative Committee to Investigate the Knox Mine Disaster, March 12, 1959: 599 and 600. Herman Zelonis was quoted by his sister, Veronica, in David Morris, "Knox Disaster Killed Deep Mining Locally," *Times Leader*, January 22, 1983: 1A and 12A. Frank Orlowski was quoted by his sister, Frances O. Sinclair, in a letter to Robert P. Wolensky on January 17, 2000. The letter is included in its entirety in chapter four. Eugene Ostrowski Sr.'s dream was described by his son, Eugene Ostrowski Jr., in a taped oral history interview, July 29, 1992, Northeastern Pennsylvania Oral and Life History Project, Center for the Small City, University of Wisconsin-Stevens Point. Excerpts from the interview with the three Ostrowski siblings are included in chapter five.

4. John Williams, Testimony, Joint Legislative Committee to Investigate the Knox Mine Disaster, March 19, 1959, 893.

5. General Assembly of the Commonwealth of Pennsylvania, *Report of the Joint Legislative Committee to Investigate the Knox Mine Disaster*, July 27, 1959, 8.

6. For a full discussion of the trials, verdicts, and sentences see Robert P. Wolensky et. al., *The Knox Mine Disaster* (Harrisburg: Pennsylvania Historical and Museum Commission, 1999, chapter four).

7. On the history of subcontracting and leasing in the northern anthracite field see Robert P. Wolensky, "The Subcontracting System and Industrial Conflict in the Northern Anthracite Coal Field," pp. 67–93 in *The Great Strike: Perspectives on the 1902 Anthracite Coal Strike* (Easton, Pa.: Canal History and Technology Press, 2002); and Robert P. Wolensky, "Competitiveness, Conflict, and Control: The Subcontracting and Leasing Systems in the Northern Anthracite Coal Field of Pennsylvania, 1900–60" (unpublished manuscript).

8. On the UAMP see Douglas K. Monroe, "John L. Lewis and the Anthracite Miners, 1926–36" (Georgetown University, unpublished doctoral dissertation, 1976). On aspects of the difficult labor-management relations in the anthracite industry see Robert Cornell, *The Anthracite Coal Strike of 1902* (Washington, D.C.: The Catholic University Press, 1957), and Perry K. Blatz, *Democratic Miners: Work And Labor Relations in the Anthracite Coal Industry, 1875–1925* (Albany: SUNY Press, 1994).

9. On anthracite's decline and economic development see Harold Landau, "Industrial Development in the Wilkes-Barre Area" (unpublished M.A. thesis, University of Scranton, 1967); Donald Miller and Richard Sharpless, *The Kingdom of Coal* (Philadelphia: University of Pennsylvania Press, 1985, chapter 9); and Robert P. Wolensky, et al., *The Knox Mine Disaster* (Harrisburg: Pennsylvania Historical and Museum Commission, 1999, chapter five). On the garment industry's contribution to economic development (as well as the decline of this industry) see Thomas Dublin, *When the Mines Closed: Stories*

*of Struggles in Hard Times* (Ithaca: Cornell University Press, 1998); and Kenneth C. Wolensky, Nicole H. Wolensky, and Robert P. Wolensky, *Fighting For the Union Label: The Women's Garment Industry and the ILGWU in Pennsylvania* (University Park: Penn State University Press, 2002).

10. The phrase "a state called anthracite" was used by Anthony F.C. Wallace, in *St. Clair: A Nineteenth-century Coal Town's Experience with a Disaster-prone Industry* (New York: Knopf, 1987, chapter 7).

## FURTHER READINGS ON THE KNOX MINE DISASTER

Beaney, Thomas M., Willard G. Ward, and John D. Edwards, *Commission of Mine Inspectors' Report on the Knox Mine Disaster,* Harrisburg: Department of Mines and Mineral Industries, April 7, 1959.

General Assembly of the Commonwealth of Pennsylvania, *Report of the Joint Legislative Committee to Investigate the Knox Mine Disaster, July 27, 1959.*

Rachunis, William and Gerald W. Fortney, *Report of Major Mine Inundation Disaster, River Slope Mine,* Washington, D.C.: U.S. Department of the Interior, Bureau of Mines, 1959.

Roberts, Ellis W., *When the Breaker Whistle Blows: Mining Disasters and Labor Leaders in the Anthracite Region,* Scranton: Anthracite Press, 1984, chapter 11.

Sporher, George A., "The Knox Mine Disaster: The Beginning of the End," *Proceedings of the Wyoming Historical and Geological Society,* 1984, 125–45.

Wolensky, Robert P. and Kenneth C. Wolensky, "Disaster—Or Murder?—In The Mines," *Pennsylvania Heritage* 24 (Spring 1998), 4–11.

## CHAPTER TWO
### RETREATS AND RESCUES

*I wore a lamp with a spotlight. You could shine a distance. I'm shining down there and I could see where we had to go down the water roofed, [went] right to the roof. I said, "There's something wrong."*
Joe Stella, surveyor, Pennsylvania Coal Company

*[Amadeo Pancotti] told me that there were five more men at the foot of the shaft. I immediately called up to some men that were along the top of that cliff, night shift mineworkers, and yelled as loud as I could to "Get rope! Get rope!"*
William Hastie, laborer, Knox Coal Company

Of the sixty-nine men who escaped from the Knox Coal Company's workings on January 22, 1959, several were still alive when our research on the disaster began in 1988. Each of the known survivors agreed to document his or her memories through an audiotaped oral history. The following excepts are supplemented by newspaper accounts, as well as the testimony of two survivors given to government investigators shortly af-

Fig. 11. Aerial view of the Knox Coal Company's operation in Port Griffith, Pennsylvania. (Courtesy of John Dziak)

ter the disaster. The stories recount the first three hours of the emergency, when forty-three workers escaped through four available exits: the River Slope, May Shaft, Hoyt Shaft, and Eagle Air Shaft (figures 11 and 12).

## Testimony: Frank Domoracki

Domoracki was the first of three men to retreat from the River Slope section of the mine. He describes the earliest moments of the tragedy in testimony before the Joint Legislative Committee to Investigate the Knox Mine Disaster. (From Frank Domoracki, Testimony before the Joint Legislative Committee of the Pennsylvania General Assembly to Investigate the Knox Mine Disaster [hereafter referred to as the Joint Legislative Committee], February 28, 1959)

> ATTORNEY JOHN FULLERTON: The roof [mine chamber ceiling] just gave way?
> FRANK DOMARACKI: And it gave way and the water come in right after it. So Fred Bohn and Jack Williams, they turned—when they saw the water, they turned around and started running up the slope.

Fig. 12. Entrance to the River Slope Mine. (Courtesy of Pennsylvania State Archives)

FULLERTON: Running up the slope to get out of the mine?

DOMARACKI: Out of the mines, that's right

FULLERTON: What did you do?

DOMARACKI: I stood down there and I started hollering to the rockmen.

FULLERTON: Below?

DOMARACKI: Down below.

FULLERTON: That had been with Williams?

DOMARACKI: Yes. I hollered to Tiny [Joseph Gizenski], "Get out. The river broke in," and I started hollering down there until the water sealed the whole slope. You couldn't see no roof, no ribs [chamber walls], or . . .

FULLERTON: When the water came in, did it go down the slope where those men were?

DOMARACKI: That's right, it sealed the whole place up. So when I seen the water sealed the whole slope, they couldn't get out, I turned around and I started running up the slope, and I passed Jack Williams and I just about got out when Fred Bohn got out and was leaning over the motor [electric underground locomotive] and I pressed him on the shoulder. . . and asked him, "How are you feeling, bud?" He told me, "If you can make it," he said, "go." So I ran in the engine house and I told the engineer—I'm winded, but I said, "Barney [Petkovyat], call [superintendent ] Bob Groves up and tell him to get everybody out of the May Shaft and shut off the power because the river broke in." And I says, "Tiny and his two buddies are drowned."

### Testimony: John "Jack" Williams

John Williams (figure 13), a native of Scotland, served as an assistant foreman at the Knox Coal Company. Previously an employee of the Pennsylvania Coal Company, his work at Knox started only three months before the disaster. On January 22, within five minutes of leaving three "rockmen," who had just completed driving a new tunnel, he witnessed the Susquehanna River's surge into the mine. If Williams had remained with the rockmen, he surely would have drowned. (From John Williams, Testimony before The Joint Legislative Committee, March 19, 1959)

## 'Roof Gave Way Like Thunder'...

"The roof gave way like thunder and lightning and the river rushed down the slope like the Niagara Falls."

So testified John Williams of 43 Swallow street, Pittston, standing at right taking the oath this morning at the Federal-State Knox mine inquiry. He presented a graphic eye-witness story of the January 22 disaster of which he was a survivor.

At extreme left, Inspector John R. Edwards, a member of the inquiry panel, administers the oath to Williams. At the inquiry table is First Assistant District Attorney Arthur Silverblatt, who is representing the county prosecutor's office.

—Wane Lance Photo

Fig. 13. John "Jack" Williams testifies before the investigative committee. (Courtesy *Times Leader Evening News*)

REPRESENTATIVE JAMES MUSTO: Would you be kind enough and let me and this Committee know where you were just before the river came in to the River Slope. . . ?

JOHN WILLIAMS: Well, I was down in the Marcy Slope [with the three rockmen]. The company miner, Fred Bohn, come down and he told me, "Jack," he says, "I heard a prop [roof support] crack up there where we are taking the pans [conveyor sections] out." So I said, "how many pans is left there to come out?" He said, "There is two." "Well," I said, "don't bother then. I will come right up."

So Fred Bohn went up and I talked to them three [rockmen] that was working in the Marcy. I said, "I am going up to see what Fred is talking about. I will be back in a couple of minutes." That was the three men that was working in the Marcy Vein. I said, "It is twenty minutes to twelve," I said, "are you going to eat your dinner?" And he [Joseph "Tiny" Gizenski] says, "No, we will load this car before we eat."

I left them and I walked up to the slope. Fred Bohn and his buddy [Frank Domoracki] was standing there when I got there. So I goes up and I stood just where I am telling you, right in front of the chamber, and I wasn't there—well, two seconds at the most, when it just let crash right ahead of where I was standing. I made a jump up to the left hand side of the rib and the water went right down past my shoulder, right down into the Marcy Vein, and all over. . . . I just got up there and I was going to go in, when that let go all at once just like a crash. It didn't give no warning at all, it just come right down in one solid crash, right down, and the water come gushing through there and across there and down into this Marcy Vein. . . .

ATTORNEY JOHN FULLERTON: How far away from you?

WILLIAMS: Well, I imagine about fifty feet. . . .

SEN. PAUL MAHADY: Isn't it true that Tiny said to the men that told you that the prop was cracking, he says, "Let us know if the river is coming in?"

WILLIAMS: I don't know. I never heard him saying that.

## Oral History: Michael "Mike" Lucas

Michael Lucas (figure 14) of Exeter had over thirty years of mining experience in hard and soft coal. A certified miner, he headed the last crew to mine in the fatal chamber. When the river broke in, he exited through the May Shaft becoming one of the first to reach the surface. His laborers, Willie Sinclair and Dan Stefanides, were not as fortunate. They perished when they took a different route. (From Michael Lucas, taped interview, June, 25, 1990, Northeastern Pennsylvania Oral and Life History Project [hereafter NPOLHP] All interviews were conducted by Robert P. Wolensky unless otherwise indicated.)

MICHAEL "MIKE" LUCAS: There were three of us in the place [mine chamber]. Sinclair was the topper, Stefanides was the laborer. . . . The river broke in [and] the story there goes, they didn't go the way I went. See, I came out of the chamber [and] went right for the [May

Fig. 14. Michael "Mike" Lucas, miner, Knox Coal Company.

Shaft] cage [elevator]. The boss had told them they could go around on top of the gob [waste rock]. The gob there had about two and a half feet clearance so that they could go around and get out over by the hospital. Well me, I didn't do that. I didn't pay that much attention to the boss.

The guy came in and told us that the water had come in [and we were to] come out as fast as we could. Yet I was stupid enough to go pick up my jacket, pick up the safety lamp and hang it on my hip, and then I went down there to the main road. It was down a little lower, we were up a little higher. So I went down there where you go out to the cage, where the damn water was running, and I figured the heck with it, I'm going right in. I waded right in with the safety lamp and everything, right into the water, and went for the cage. Then after I left my buddies, they were going around that gob to get out over there in that slope opening. That was a long ways . . . A long time before [earlier], he [foreman Frank Handley] told us how to get out of there just in case this and that and the other. See, he sort of suspected that that might happen. That's what I think anyway. He sort of suspected what happened so he gave us advice as to how we would get out. Most of the time, I don't know, I just didn't believe all of the stories that they'd tell me because they'd be telling you things to suit their purpose, you know. They had to go within the law and this other stuff. . . .

Billy was the topper. He could have, what the hell, he could have gone [out] right after he told me. See the coal would come down and when the car got filled up he'd pick some chunks and put them around so there would be more coal in the car. He was the topper. He wasn't in the place, he was down where the car was in the main road. Well, he waited for Danny because him and Danny were pretty thick, you know, good friends. As far as that goes, we were all friendly. . . .

ROBERT WOLENSKY: Did you tell them to follow you?

LUCAS: No, I didn't because it was everybody's choice, see. I could have been wrong just the same as they were. But whoever made the right choice was the one [who made it out]. . . .

When I got out there on the straight and ready to go for the cage, I saw Zack Zakseski. He comes walking out there behind me. I went straight for the cage. Zack Zakseski and me got on and we got out. The water was running for the cage already because it was a little deeper there. I got on the cage and rung the bell and had to wait a while. I was sort of partly scared. I thought, "Jesus, this guy ain't going to lift us up." He [eventually] lifted us up. . . .

[When I got out] I went straight over to the super's office and I told Groves I said, "Bob, you got to get them out." I told him that. I was sort of then, and even after that, I was sort of, let's say upset. I came home some way or another. I remember that my wife and me went back over there. There was a lot of people around there, you know. . .

After this happened I went up to tell Billy's wife that I felt sorry and she really laid into me. I didn't even get into the house she started

hollering at me. "How come you're here and they're not?" and so forth and so on. I didn't even get a chance to talk. She was sort of like that and it really upset me. See, my doctor way back already told me not to get excited or anything like that, for me to keep it cool.
WOLENSKY: Did you ever talk to Danny Stefanides' wife?
LUCAS: I don't know if I did or I didn't. I might have told Danny's brother to tell her. I don't know what I did in that case. I try not to let it get on my nerves. . . . I felt pretty darn bad. I still feel bad about it. They didn't go the way I went, they went the other way. They made the wrong decision. But after I got out I felt like if I could have taken them both by the neck and drug them with me instead of letting them go the way they went, because I liked them guys. They were my real buddies, you know, good friends. But even today when I cross the river I'd be saying, "Dirty ol' river" . . . .

It broke through in a chamber where we worked about eight days before that. We were working in another chamber then already. There were lots of droppers [water] dropping from the ceiling. It was wet. It was real wet in there. We'd come out wet. The coal was good. If I'm not mistaken we were working what was called the Big [Pittston] Vein. We were taking pillars. The place was all mined [out] and these were part of what was left. We were going around the end and taking them out. I forget what the hell it was now. I think we just drove into it. They were getting coal where they could get it—the easier to make money the best.

## Oral History: Chester "Chet" Dunn

A certified miner, Chester "Chet" Dunn (figure 15) of Hilldale, Plains Township, was working as a "motorman," or electric locomotive operator, in the May Shaft section of the mine, delivering "empties" to the workers and hauling "loadeds" to the surface. After receiving word of the breach, he ventured deep into the pit to warn others. Contrary to the general view, Mr. Dunn maintains that when superintendent Robert Groves called underground ordering the evacuation, he indicated that the river had broken through. Mr. Dunn escaped through the Hoyt Shaft

Fig. 15. Chester "Chet" Dunn, motor runner and miner, Knox Coal Company.

with two other men. (From Chester Dunn, taped interview, August 4, 1989, NPOLHP)

> I was in there [motor barn] waiting for the miners to load their coal. In the meantime the telephone rang, but each one in that shop has a different ring, one ring [for another person], two rings for me. This time the telephone rang steady which meant there was no ring for anybody, and it kept [ringing]. It rang for about five minutes. So I picked the phone up and it was the superintendent of the coal company, Bob Groves.
> 
> He asked me who was on the line and I told him, "This is Chet." "Well," he said, "Listen, the river busted through at the River Slope. Would you please go down the lower section of the mines and notify anybody you could see?" Well, I wouldn't take it that it was that seri-

ous, so I said, "Sure." I went down and every miner that I spoke to I told them, "Listen, you are familiar with River Slope." I said, "I was notified by Bob Groves that the river busted through. So everybody as soon as possible get out." At that point I said [to myself], "I'll have to go up the top lift [vein] and speak to the other miners."

If I didn't answer that phone, I'll tell you none of us would be here. There was nobody [in that section who] would know the river came in. At the meantime, when I was going up the slope, I ran into the main electrician in the mine, Fred Cecconi. He asked me where I was going, and I said, "Fred, I got word from Bob Groves that the river busted through the River Slope." "Oh my God," he said, "I didn't know anything about it." I said, "Where're you going?" "Well," he said, "I'm going out [to the surface]. I'll go throw the main switch," the main electric switch. There's another incident. If I didn't run into him we wouldn't even be here because we would all have been electrocuted. When the water got to a certain level it would have hit the main power and everybody would have been electrocuted because we were in water. We were ground, see? So when I saw the lights go out I figured, "Thank God, Freddy got out. He shut the power off."

Then I started to come back to try to get out to the May Shaft. When I got up so far, I couldn't make it because the water was up to my chest. . . . [It was] very loud, just like thunder. When it came in, with full force, it was just like an avalanche. I couldn't say how fast it was moving but it came in fast. It spread all over the section, all over the mine. No matter where you look, you saw water seeping in. I threw my jacket off and [took] everything off to try to get [out] as fast as possible. . . . What was going through my mind—to try to get out because it was my last chance.

There was only one way, I said, that I can try. So I went down to what was called the Baloga Slope and I ran into two men, Joe Shane and his brother George. They were from Inkerman. Joe asked me where I was going. I said, "Well I'm going to try to get out to the main [Hoyt] Shaft. If we can only make it, Joe," I said, "we'd be okay." The three of us went down [in that direction]. But in the meantime while we were going the water was following us. I was up to my neck because I was the shortest guy and the other guys were a little taller. Finally, we got to the elevator and we rang the bell. I'd say we walked about ten minutes. We got into that elevator and went up. When he left us off I told Freddy Cecconi, "Look, drop it down, maybe somebody else is fortunate enough to get there."

See, that's why he [Cecconi] went out—to run the elevator. When he started to drop it down, the elevator got to about fifty feet and the cable hit the ground because the pressure of the water pushed the elevator off and it jammed. It wouldn't go anywhere. Nobody else came out of the [Hoyt] Shaft.

[When I got outside] I blessed myself and this is why my name

Fig. 16. Frank Handley, foreman, Knox Coal Company.

was never mentioned because I walked from Port Blanchard to Port Griffith to get my car because that was the shaft that I went down in the morning. I had to walk about five minutes. In fact, the three of us did. When I got in my car, all my clothes were wet. I came home. I didn't want to be bothered by cameramen or anyone. I just came home and the first thing I did when I came home—my grandson was six months old. He lived right up the road here. I grabbed my grandson and I hugged him and said, "Honey, your grandfather made it." He was only six months old and at the time I was forty-one years old. I never went down the mines again.

### Oral History: Frank Handley

Frank Handley (figure 16) of Kingston, a veteran mine foreman, had asked to be transferred to May Shaft section because he did not approve of the mining practices in the River Slope. Supervisors granted the transfer which is why Handley was not indicted by the authorities. When the river rushed underground, he led twelve men to the surface through the Hoyt Shaft. Like Chester Dunn he remembers that superintendent Groves' evacuation order included word that the river had pierced the mine. (From Frank Handley, taped interview, December 10, 1988, NPOLHP)

> I answered the phone and I got a call from the superintendent, his name was Bob Groves. He said, "Frank, the river broke in at River Slope; get all the men out." Well, boy! It took us about fifteen or twenty minutes to get these men together, and I said to them, "Our chance is maybe to go to the Ewen [May] Shaft." Pennsylvania Coal Company had an engineer at that shaft and it might have been maybe quarter of a mile [away]. It wasn't far. Otherwise if we had to walk up this slope and out to the main entry, it'd take us maybe three times longer.
>
> Well, when we started it struck me that the boss at the Pennsylvania Coal Company told me, he said, "Frank, we have an engineer there every day but Thursday." That's the day off for the engineer and this was a Thursday morning. Jesus Christ! So anyhow, I rang outside and I told the superintendent, I said, "We're gonna go to the foot of Hoyt Shaft, get an engineer there." He said, "Okay, Frank." And we

started back. It was quite a day. I told them then, I said, "They're gonna get an engineer." The men, you know, get a little jittery. Before we got there, I'd say the water was maybe up to my knees.

When we got to the shaft, you'd have to ring the bell and they'd answer, see. I rang the bell, two bells. The carriage came down. Then when you get the carriage, you ring one bell and get on. That was our signal. You put the other men on first, see. There was eleven [twelve][2] of us. They all got on. When that carriage [started] the water was about up to my waist.

I opened the gate and they were all in and I said to one good sensible fellow there, his name was Joe Kopcza from Wyoming, I said, "Joe get in." I said, "We won't back track the signal. When I get in I'm just gonna hold and give them one [ring] and that's it." They got in and I rang the bell, one bell, and jumped in. And that Joe I'll never forget, he helped me. He was worried too. The water was just about waist high.[3]

I often think though, now that we're talking, when we landed on top the superintendent [Groves] said, "Frank, is there a chance in the world that there'd be somebody there at the bottom?" "Hey man," I said, "No one's ever gonna roam around there." He said, "Do you think it'd be worth a chance to go down and see?" I said, "Well, it's okay with me but what I left was nothing." He went to the engineer and he said, "We're gonna go down." He got on with two or three fellows and he rang that bell twice to drop them down with the understanding that they would stop above the vein and look and see, [then] ring the bell to haul them outside. But they never could move the carriage because the other one wouldn't come up, do you follow me? See [there were two elevators] one's going up and one's going down. But as soon as we got off that must have been the end of it. They never could get the other one up. We were the last ones out of that exit.

### Oral History: Thomas "Tommy" Burns

Thomas Burns (figure 17), a laborer from Pittston, discusses the difficulties associated with his egress as a member of Frank Handley's group, as well as the nearly calamitous descent of three men into the May Shaft to check for stragglers. He also recalls his good friend and disaster victim, William Sinclair. (From Thomas Burns, taped interview, August 8, 1989, NPOLHP)

THOMAS "TOMMY" BURNS: We were working, six or seven of us. We were setting up what is known as a "jalopy," a chain [loading] machine that they used when they were going to drive chambers downhill. Incidentally, working with us that day was the chief electrician of the Knox Coal Company [Fred Cecconi]. We got everything done, he was done with his work, and he was connecting the setup. He left the area to go up on top, meaning the May Shaft. We continued to work getting pans connected and doing all that had to be done, cleaning up.

We got the word from a company miner that was working along the main road timbering. It was approximately 11:15 a.m. or 11:30 a.m. He said that we were to get out of the mine. We started out, six of us I think there was. We walked and came to a slope, the name of which I don't remember right now, and we met the mine foreman, Frank Handley. Frank Handley is one of the most courageous persons that I can lay it on to. He was waiting for us knowing at the time that the river was in, yet he waited at the foot of this [inside] slope until all the men got accumulated. Then he told us that he was going to take us through the old workings to the Hoyt Shaft. . . .

Mr. Handley was ringing the bell for the cage to be lowered in the Hoyt Shaft. We stayed there about ten minutes and there was no answer to the bell. There were two bells to send the cage down and the engineer up there [on the surface] gives you one bell noting that he's sending the cage down. Well, there was no answer.

Mr. Handley said we may as well go back and try to get up to the River Slope. So we started and Mr. Handley stayed at the cage, at the foot of the cage, and kept pressing that bell. He finally got an answer. That's when he called us back. By this time there was thirteen [twelve] of us counting Mr. Handley. He called us back and the cage came down and we all got on the cage. Usually they only leave ten people on the cage. Everybody crowded on the cage and Mr. Handley gave the signal. The water level was rising.

We got half way up and the cage stopped! And then it started

Fig. 17. Thomas "Tommy" Burns, laborer, Knox Coal Company.

again. Well, we found out later—some of the things you find out later are really unbelievable. . . . [that] the rightful shaft engineer had the day off . . . [so] them cages were suppose to be out of commission. But Freddy Cecconi was sent over to run the engine that hoisted and lowered the cages. So he got over there and answered our bell. Well, he got our cage half way up in the shaft when the rightful engineer, Dominick Ochipini [?], got there. Freddy had him take over to complete the operation of raising us out of the mines. We didn't know that, of course, we just wondered what the hell he stopped for!

We got off the cage when we got to the top still not knowing what was going on. We still didn't know. I don't think Freddy was there, he had gone back up to May Shaft to see if he could do anything [there]. Anyway, Frank Handley the mine foreman, we asked him what the hell is going on. He said, "You'll find out." Well, this shaft had two cages in it, as you raise one the other lowers. That [other] cage that had descended while he was raising us out of the mines, it was down below the landing a little bit, and Dominick wanted to raise it up even with the landing in case there were other men that were gonna come through the workings. It could be there for them. Well, he never did raise it. The water was in it. That cage is still in the mines.

We didn't know anything was wrong until we went up to the shifting shanty [miners' wash-up shanty] and then they told us the water had broken in. Then I changed and I called my wife up. God, I loved her. She didn't even know that the water had broken in. See, my miner and me went home. I lived here. He lived right up the street. I wanted to get back down there [River Slope] to see if there was anything I could do to help these people that were still in the mines. So I called my wife from one of the houses in Port Griffith and told her I said, "Now don't worry I'm alright." She said, "Well, why should I worry?" Why should I worry, she says! I said, "Well, the river just poured into the mines and there's twelve or thirteen men that lost their lives already." This shook her up. When I told her I was going back to work she said, "My God, Tom what are you gonna do?" I said, "I wonder if I can help them down there."

So I went down [to the River Slope] and I met Cab [Knox employee Anthony Waitcavage] who was doing company work with his brother loading coal. We stood around there to see if there was anything we could do to help. We stayed there all night sitting on the river bank. By this time the men had come up through the Eagle Shaft, which I had seen thousands of times as I was growing up. . . .
ROBERT WOLENSKY: Did you know any of the men who were killed?
BURNS: Just one. Billy Sinclair. Billy Sinclair was one of the finest persons in the world. He was a lovable man, lovable man. We use to ride [together] when he was on the day shift. I rode home with him daily and we usually stopped and had a couple [drinks]. . . . He had a sense of humor that was nineteen miles wide and there was still a little Scottish [brogue] to hang there, you know. Every time I worked with him he was a darn good worker. He was an honest worker; yes

he was. He was what they called a topper.

From what I understand, Billy Sinclair worked approximately 100 to 150 feet from the foot of the May Shaft. Now from what I understand, how true it is I don't know, but when Freddy the electrician foreman was going out of the mines, he seen Billy and he said to Billy, "Call your men, your miner and laborer, and tell them to get out of the mines." He never told him why, just told him to tell big Mike [Lucas] and Danny [Stefanides] to get out of the mines, get out of the mines. I don't know whether Billy went in or whether he gave them the signal by stopping the jalopy. But anyway the miner [Lucas] got out of the mines. He was a big six foot two or three, a raw individual. Billy and Danny didn't make it but Mike made it. People talked to Mike afterwards and I heard that he had a struggle. He was holding onto a trolley line to pull himself to the cage, but he made it.

Just thinking about Jimmy Jamieson, the [assistant] foreman in our section—he came up the Hoyt Shaft with us. During the period while the cage was operable in the May Shaft, they were able to lower the cage. Bob Groves the superintendent, a man by the name of Tom [Watkins] who worked timbering the shafts, and Jimmy Jamieson who just came out of the mines and knew that the river was in, got on the cage. The three of them go down. Bob said in his Scottish brogue, "I wanna see." He wanted to see what was going on. They went down in that cage and they were down there maybe about two minutes and you could hear them all hollering, "Take us up, take us up." They couldn't take them up because a fuse had blown. I'm talking serious. The bosses were hollering down that a fuse had blown.

In the meantime, them not knowing what's going on, there was on opening of about yea much between the cage and the shaft wall. They crawled between the cage and the wall and got on top of the cage [because the water was rising]. Now they did this not knowing what was going on up on top. In fact, what was going on up there was Freddy Cecconi was looking for a fuse. He found the fuse and inserted it and immediately the cage moved. Now just place yourself in the middle of those guys—if one of them was still climbing between the wall and that cage it would have cut him in half.

Fig. 18. Joseph "Sheriff" Kopcza, miner, Knox Coal Company.

## ORAL HISTORY: JOSEPH "SHERIFF" KOPCZA

Joseph Kopcza (figure 18), a miner from Wyoming, exited at the Hoyt Shaft. He recounts the tense moments surrounding the elevator ride to the surface with Frank Handley. (From Joseph Kopcza, taped interview, December 21, 1988, NPOLHP)

JOSEPH "SHERIFF" KOPCZA: I had four men with me. We were sent [underground] this day on the twenty-second of January to put up a machine near the Wyoming Bridge or somewhere back up in there, but away from the barrier pillars. A barrier protects the water from going any further. They got permission to go down in there, I guess. Handley knew more about it than I did. So we were setting it up and the boss came down, the superintendent Bob Groves. Handley was with him and [so was] the electrician foreman, Freddy Cecconi. They looked the job over and I'd say about 11:00 a.m. they left.

After they left I told my fellas, "Well look, there's no use starting to work now. In a half an hour we'll be having lunch, so let's sit down, we'll eat our lunch, and we'll start in and finish the job before we go home." They all agreed.

Soon after we sat down I heard this voice hollering, "Hey Joe. Hey Joe." I said, "Listen fellas somebody is running." In the mines when a fella's running you can hear the taps of his feet on the bottom. With the air and everything so quiet I heard this thumping and hollering. I ran down and I said, "What's the matter?" This fella's name was Joe Nedalski [?] from Avoca. He said, "get your men out of here." He didn't know what was the matter. He just told me to get out.

So I told the fellas, "Well boys, you'll eat your lunch outside," kidding around, you know. We kept going and going and then I came to a machine, a shaker chute. We had a separate machine there and in case something went wrong you would always put a man on there and he could finish his day's work. I threw the switch—no power. I said, "Uh oh."

Down in there with no air it could fill up with gas fast. I said, "Fellas, we have to get moving out of here because this is a low spot and it fills up with gas. You never know what's going to happen." I went up further and I met Handley, Jamieson, Shaney [George Shane], and these two, Nedalski [?] and [Robert] Robshaw, were with them. I said, "What's the matter?" Handley said he didn't know. But [when] I saw the water coming down the River Slope I figured it was a fella working up on top the slope, Zack [Zakseski] was his name. I thought he busted into a body of water. Sometimes in the mines you run into those things when the water has been accumulating and then you bust in, it'll come out. . . . There was a lot of water up there and it was direct almost with the slope. But jeez, when I got closer to the slope, I saw the water coming down and I said, "That can't be a body of water because it's not rusty." It was clear and that gave me an idea [that it was the river.] I didn't say anything to anybody because if you do you'd scare everybody. But Jesus, the water was coming faster and faster and then I saw a piece of ice. I said, "Holy God, where is this [coming from]?"

We were walking in circles half of the time—you go up, you go here, where you can't get out you go further, you come back. What we were trying to do is follow the old roads. They were going into different working places. Finally we had an idea. This plane gave us an idea. I heard about this plane but I never worked in that section and it was something new to me. But I heard fellas talking about it. They used to pull the cars up from the shaft and drop down the "loadeds" and the empties, to branch them around. They didn't use motors [in those days], but [instead used] this plane. [It had an] electric engine. I'm saying it was just God's luck that we found it. There was no other way out for us.

Frank Handley suggested that "we will have go for high grounds." Handley was a smart mining man. He was a good sense man. He didn't get excited and he meant a lot to us there, Frank Handley did, because he kept us even. He kept everybody well balanced. I said, "Frank, how are we going to go for high grounds [if] we can't get up that way?" Nobody knew anything. I am saying it's God's luck that we found this plane and came out of the Hoyt Shaft.

See, when we got up there was no water because of this ditch. That culvert saved us because if the water wasn't going in there it would have filled up and come up to the plane. . . . So we figured we couldn't get up the slope and the only place we could get out, I was telling Handley, was the shaft over here, Hoyt Shaft. I never worked in the Hoyt Shaft area. I didn't know but I heard them talking about a plane they used to hoist the cars, take them up and bring them down. This all came into my mind and I told Handley, "There's a plane here somewhere and the Hoyt Shaft is this way." We found it! Just God's luck that we found it.

We went down there and they had a water tunnel right in the shaft. About twenty-five feet away from this water tunnel we found the shaft. The water was coming down this tunnel or culvert, and it was about maybe five or six feet wide, about eight or nine feet deep. The culvert only had about a foot of air going through. The water was going all the way into the Schooley Shaft probably. We jumped over that water tunnel. I think that's what saved us.

Freddy Cecconi knew where we were at. He figured that's the only place we can come out. That day the engineer was off. Thursday they had no engineer there. They heard us. We were hollering and banging and everything to try to get the carriage to come down and take us up.

God was with us then too. Cecconi put the carriage right on the landing, perfect landing. He didn't know where he was at because everything was all shorted out, buzzing and ringing. It stopped. We all got on. It wouldn't move, no how! Well, there was a wooden gate there and Handley said, "I am going to get off and tell them to pull us up." I said, "Handley, don't get off." I was holding him because once this [elevator] would move Handley would never get on. He was stuck, he was done, and I wouldn't let him go. I am holding on to him, and holding. I said, "Don't you go." He said, "Joe, I'm still your boss." I had no choice, you know.

So we broke the gate and he got off. He got on the telephone. "Pull away in one minute," he told him. He hustled right on and I am telling you I don't think he quite turned around when the carriage started going up. It went about maybe twenty-five feet or thirty and stopped! I said, "Oh my God, now what?" The engineer came in; they switch engineers. We came up and that carriage never moved no more. . . . See, the one [elevator] went down the bottom [as we came up] and probably the guides got twisted and they got off the track. It jammed and they couldn't get it up or down. It only made one trip and if there were any men down there they'd never get up because the carriage couldn't work. I don't think [there were any men down there] but I'm just saying if there were they were done because the carriage was stuck.

ROBERT WOLENSKY: Was there water in the cage?

KOPCZA: It was starting to come up. It was on the floor of the carriage a little bit but not much. The Red Ash Vein [the lowest vein] was getting filled. We were fortunate. If we would have been there maybe another half hour [we wouldn't have made it]. . . .

WOLENSKY: Who were the men that came up in the cage with you besides Handley?

KOPCZA: There's Johnny Stupak, Johnny Smutko [?], [James]

Jamieson, Joe Nedalski [?], a fellow named Robshaw, Leo Vella, and Tommy Burns.

I picked up my clothes and got the heck out of there. I didn't want no part of it. The next day I was called by state mine inspector, Dan Connelly, to go to work [to help at the mine]. I said, "I am not fit to go to work." He said, "You come here even if you have to just sit down and talk. Get your bearings." It took me quite a while to get back because you're in shock. There's a lot of people like that but they didn't know the difference. We had one fella, Leo Vella. Frank Handley told me, "Joe watch him [because] his eyes are glassy." We thought he was going to pass out [as we were exiting]. I was right on top of him. We'd put him out, knock him out, so as not to flare the other guys up. We'd have to do it, but he held on and all of us got out safely. . . .

I tell you a person does everything [to get out of the mine]. When I seen rocks coming down like whales in the ocean—big rocks—then ice coming down, you do everything. You do whatever comes in your mind. It wasn't that I was scared. I think things were running through my mind so much you didn't think about being scared. You were thinking about getting out!

## Oral History: Anthony "Tony" Remus

Anthony Remus (figure 19) followed in his father's footsteps by working at the Pennsylvania Coal Company's Ewen Colliery, beginning at age fourteen as a nipper and later moving up to a laborer. He began at the Knox Coal Company a few years before the disaster. He joined Frank Handley at the Hoyt Shaft for the trip to the surface. (From Anthony Remus, taped interview, December 28, 1988, NPOLHP)

> Well, I went to work [and] registered. Our machine was broken. The night shift man broke it and they put us in a different place. I'm lucky they did put us in there or otherwise I'd be still swimming.
>
> Jimmy Jamieson, the assistant mine foreman came up. He hollered into the chambers, "Everybody out." We were taking our time. This miner forgot his safety lamp, he went back for it. . . . There was one guy, his name was Baloga who from what I heard said, "I'm going to go put my clothes away." He never came out. Now if he knew that

Fig. 19. Anthony "Tony" Remus, laborer, Knox Coal Company.

the river came in, he might be outside today. . . .

We were walking out and all of a sudden, water. . . . I said, "There's water! There was no water here when we came in." . . . We turned around and there was water up to our knees and then Frank Handley said, "This way," he said, "I know the closest way [out]." I guess he knew where he was going but we didn't know it. The people in front of me were running. I wondered, "Oh, what the dickens," so I turned around and I ran. I took my water bottle and everything. You could already see the water coming. I'm thinking it could be the river. That kind of shocked us. . . . I couldn't hear any noise. You weren't interested in anything. The only thing you wanted, get out, that's the only thing. Already the shock was in you. You know what I mean, the scariness? By just thinking of it, you know, when the water started coming. . . .

There were two ways to get out, the May Shaft and the Hoyt Shaft. But the May Shaft was getting the water, they said. My miner went down that way and the [other] labor went down that way [and they got out]. They were in water up to their chests. They found the engineer to run the engine. He went down there and picked [lifted] them up. But Frank Handley said, "This way." I don't know if he meant that we were safer or closer. . . .

Then I got on the carriage. Oh, there were quite a few of us got on the carriage. The water was up to my knees. We got up [surfaced] then found out it was the river. The carriage went [back down] but never came back up again. [We] made the last carriage. . . . They went down with the bucket the next day. They couldn't see anything, not while the water was there. . .

I went to the shifting shanty and [then] home. When I saw my brother everything was okay. He was a miner. I came home and told my wife what happened. I said, "I'm going down to the beer garden." I went down to the beer garden and got loaded. It was tough.

### Oral History: Joseph "Joe" Kaloge

Joe Kaloge worked as a laborer in the same crew as Anthony Remus. He emerged from the same exit at the same time. (From Joseph Kaloge, taped interview conducted by Ellis Roberts, March 6, 1979, NPOLHP)

Jimmy Jamieson came in and he said, "Everybody out." Seems like we didn't know what was coming in or anything because we took our time. Then on our way, when we heard the water roaring, we figured [that the river came in]. . . .

See, this way you go down to the May Shaft and that way you go down the Hoyt Shaft. [Frank] Handley said, "Come this way." I took my jacket off. I knew to go this way. I figured this way was closer. We made that last carriage. . . .

While we were working the roof was like cracked and water started seeping through. But we never thought the river would break through. We were working one level down. The water was above us. Where that water was coming from it started to drip like it always does in the mines. But this was like cracks, seams. The water was coming down and you just wondered where it was coming from.

If we worked in our own chamber, well, I'd be still there. Our jalopy broke that night and they put us about a half a mile away from where we got out. We had been driving a place that was, say, three miles away from the exit.

### ORAL HISTORY: MICHAEL "MIKE" OBSITOS

Mike Obsitos (figure 20) of Swoyersville had been working for the Knox Coal Company for three years before the fatal day in 1959. Speaking in broken English (he was born in the U.S. but his Slovak parents spoke the native language at home), Mr. Obsitos vividly describes his passage through the Hoyt Shaft. (From Michael Obsitos, taped interview, December 8, 1988, NPOLHP)

MICHAEL "MIKE" OBSITOS: [Motorman] come in. That was about eleven o'clock in the morning. . . . He says, "Hurry up go home quick." He don't say river come in. . . . We left the [lunch] pails, everything. Coal over there—we never finish it. We don't finish our cars like them guys was finishing [theirs]—they still over there [i.e., the victims are still in the mine]. We come out on the [main] road. Ice was flowing down the hill. Ice right behind us. We come by where Danny Stefanides was working. Water was coming through already. We grab the wires, electric cables, and go over. . . .

I was about three thousand feet from the [elevator]. We was walking on side roads first. We ran. Guy was sixty years old, he beat us out. He beat us out running! And nobody have pails. Everybody throw everything, just go. More than half a mile we was [away from the elevator]. [We could see] ice when we come out from our road this

Fig. 20. Michael "Mike" Obsitos, miner, Knox Coal Company.

way. Go straight. We see ice going like down a hill. All ice. And motor right behind us. . . . The motorman wants to give us ride on motor, only when he see that water, he left the motor and he start running with us. Nobody was riding motor and water was pushing right behind us. It [water] was about up to knees there by the shaft. . . .

We come by the shaft and Jesus, we're blowing [winded]. They put jumpers and take her [elevator] out. Nobody else come out no more. Last ones to come out. That cage, fuse blow. They jump it with wires and they pull us out. Water was going under there.

ROBERT WOLENSKY: So you made it by just a few minutes?

OBSITOS: Yeah. Nobody else come out. We watch over there, wait in the shiftin' shanty. Was pretty cold that day and the shiftin' shanty was nice warm. Everybody watch [but] nobody come out no more. . . .

WOLENSKY: How close was Dan Stefanides to the shaft?

OBSITOS: Thirty yards, most. . . . . When we come out Mike Lucas was out already. Sure, he was only thirty yards from shaft! He was out. First thing we ask him was, "[Where is] Danny?" He says, "I don't know. . . ." We don't know what's happen. When we come out, then we know river come in. Gee, like Niagara Falls the river was coming down. . . .

WOLENSKY: Did you ever expect anything like this?

OBSITOS: Yea. We know guys was working under there [river]. They have raincoats on and it was dripping water. It was leaking, river. And inspectors was over there. They didn't say nothing. They know it gonna [break]. They musta know. Was nice coal over there, maybe twelve feet high. When you load thirty-five cars, that's lots of coal. Three men! They have a Joy loader [mechanical loader] over there. Gotta like arms. You just make the coal and swish 'em in and it dump in the car. . . .

Miner boss was Handley. I work for him over here at Harry E. [Colliery]. That's why I got job over there [at Knox]. Was a good job. Not dusty. Handley says, "Come on, I got job for you." We went—Mike Hanusan and Pete Slempa. We all got out [escaped from Knox] together. We was home by two o'clock. [We went to] Dicko's [bar in Swoyersville]—me, Pete Slempa, Mike Hanusa. Everybody [came in the bar to see us]. They hear on the radio, you know, and they was thinking, we over there too.

WOLENSKY: The people thought you might have been trapped?

OBSITOS: Yeah. . . . Then in the night me and my wife went over there by the mines. [We] see everybody come out [of Eagle Air Shaft]. They was marking who come out, who not, who down in mines. They know who was working day shift.

Fig. 21. Pacifico "Joe" Stella, surveyor, Pennsylvania Coal Company, rescued from the Eagle Air Shaft. (Courtesy of Stephen Lukasik.)

## Oral History: Pacifico "Joe" Stella

Joe Stella (figure 21), Pittston Township, began the morning of January 22, 1959 by surveying sections of the River Slope for his employer, the Pennsylvania Coal Company. He became one of the heroes of the disaster as he safely led six men in a two-hour trek to the Eagle Air Shaft where they eventually climbed to the surface. Before he exited, Stella returned into the mine with two colleagues to search for a lost party of twenty-six led by assistant foreman Myron Thomas. (From Joe Stella, taped interview, November 1, 1988, NPOLHP)

PACIFICO "JOE" STELLA: What happened when I was in there that day, I was making an inspection with Myron Thomas of what we called the substation area.... While I'm waiting for him, I was putting my map up to date of what we inspected in his working chambers. As he's having his sandwich we could hear this motor that transports the loaded cars to the shaft and would bring empties back for the men to load.

Myron Thomas tells me, "I don't like the sound of that motor. It's really barreling in here. It [usually] doesn't come in that fast, and he's empty." So Myron was thinking that probably somebody was hurt or something, and they called for the motor because we weren't in the office, we were out inspecting his section. The motor runner came right up to the door where we were standing. He said, "Myron, we just got word for everybody to get out of the mines." Myron told the motor runner, "You go back out and go down to the Marcy and get those men out. Me and Joe will go back and get those guys out of the substation here."

ROBERT WOLENSKY: Did he say why you had to get out?
STELLA: No. He just said everybody get out of the mines as fast as you could. Back in we go, Myron and me. We get all the men. He had about three or four chambers working in there, which would have amounted to about ten to twelve men. We got them all out and we met where they would go down to the Marcy. We waited there until this guy came up with the men from the Marcy. Then we said, "Okay, well then, let's start heading out."

We started to head out [toward] the main exit [May Shaft], the way we came in. I remember I was on the lead and I started stepping in water. You couldn't tell it was water because water that rises slowly has a film of dust over the top of it [so] that you can't tell there's water there until you start stepping in it. I said, "Look, there's something wrong here. I'm stepping in water." As I'm going ahead the water's getting higher. I didn't go any further than about a foot deep. My boots weren't too high. I said, "This is all water."

I wore a lamp with a spotlight. You could shine a distance. I'm shining down there and I could see where we had to go down the water roofed, [went] right to the roof. I said, "There's something wrong." We didn't know where the water was coming from. Myron and I decided the closest place would be to get out through the River Slope. I said, "Yeah, go ahead." We start heading toward the River Slope. As we're going toward the River Slope the pressure of the water was getting terrific. You couldn't get into the water. If you did you were gone [because] the current was so swift. And the big chunks of ice! They tell me the ice that day was about sixteen or eighteen inches thick. Them chunks of ice were coming down and taking those mine cars and just pushing 'em. What saved us is that the workings were going up on a grade. You had the main heading, [it] was like in the basin, and then they mined to the right and to the left which went uphill just like a "V". . . . What saved us is we could stay away from that fast water and cross from one chamber to the other by staying up in the higher part, up in the higher elevation.

We continued toward River Slope. . . . As we were getting closer to River Slope, boy the noise was so tremendous with the breaking of timbers and ice and everything, you couldn't hear yourself talk. You could only motion with your light, wave to somebody if you want them to come over or either wave it across for somebody to stop. So finally, as we were getting closer to the River Slope, the chambers from this ten- to fifteen-degree angle, they started to level off down to ten, five [degrees] and then the water comes right up to the roof and we can't go any further. Myron and I decided the only chance we had was to head to the old Eagle Air Shaft.

I knew about that shaft because five or six years before that I made a survey with the surveying corps. We went in and we surveyed all those old mine workings. The Pennsylvania Coal Company drained the water out. They were all filled with water. The Pennsylvania Coal Company gave permission to Knox to advance those old workings slowly because they didn't know exactly where they were situated. They were never surveyed. . . . When we surveyed that and went in and plotted our maps and made new maps, it gave them a good idea of how much additional mining they could have done. It even shows on the map all the workings that we surveyed up there. . . . That's how I knew that shaft was there. I'll never forget it because it was a real cold day in January [when we surveyed]. The air was pulling in. Oh, we were freezing. The shaft was so close to the railroad track that if a train was driving we could hear it because it was about forty-five or fifty feet to the outside. You could look out there [and see] if it was snowing or whatever.

That day, the Eagle Shaft was the only thing that was left. Lot of them were saying, "Let's go down the Marcy Slope, we'll go out through the Schooley," that would be over on the west side. I said, "No, why should we go down lower, we know that the river is coming in from the River Slope and to go to the Schooley we would have to go down the slope to the Marcy, the next vein [down]."

What happened, after we turned around and Myron Thomas and myself decided to head from the River Slope to the Eagle Air Shaft, we separated because I had four or five or six old fellows. I guess [Thomas' group'] was going a little bit too fast for them because they were up in age, maybe well in their sixties. That's all I could hear, "Joe, wait for me, wait for me." I [earlier] told Myron, "You know where the shaft is, you take the lead and I'll stay in the back with these older fellows." That's how we split up. They took off. I said, "That's OK, we'll meet them up at the air shaft."

As I'm going along I find a couple of more men. I said, "What's the matter?" They said, "We couldn't keep up with them," so I said, "Well come on, I know where the shaft is; we'll see if we can make it up there." I had a map with me. I'd check myself every now and then and I'd find what we'd call "stations" on the map. They show on the map and then they show on the [roof of the] mines. They would be control points to make sure that I'm going in the right direction.

We were about seven [in my group]. All together we were about thirty-four [thirty-three] people, the miners and laborers. I get up there [Eagle Shaft] and I was surprised, because, like I say, about five years before that you could see right outside that shaft. When I got up there that day with the men, you couldn't see outside at all. They had it filled in! I told them, "This is the shaft here because you could feel the air pulling." The ventilation in your mines—you have fans that draw, not push, air. So any opening to the surface would draw in. It's easier to draw the air than it is to push and you can do more ventilating that way. What really surprised me when I got up there—there was nobody there. . . . I said, "We'll go and look for some tools [to dig away the debris]. I said, "There's a couple of chambers close by here that were working. We have to go and bust the miners' toolboxes."

I told the other guys, "Look-it, while you're staying here, pull this stuff down," because there was all big boulders and dirt and garbage and everything. I said, "We're going to go back looking for tools." We went back, me and Jerome Stuccio and Jimmy LaFratte—they said, "We'll come with you, Joe." We only had to go back maybe five hundred feet to those toolboxes where there were live chambers working. What did we do? We went all the way back—a couple thousand feet—

all the way back to the Marcy Slope again to see if we could find the other crew.

When we got there, oh the water was going down that [Marcy] Slope. Jerome said, "Let's go down part way, maybe that's the way they went." I said, "Listen Jerome, if they went down and tried to get out through the Schooley, they're done." I said, "Look at the way that water is going down there. What chance do you stand?" I said, "I'm not going there. You wanna go, you go! I'm going back, we'll break couple of the miner's boxes and carry whatever tools we can."

Back we come. We broke couple tool boxes and we carried whatever we could carry—picks, shovels, dynamite, firing line, a battery to fire the dynamite—whatever we could carry, the three of us. We were coming with those tools and making our way back to the air shaft. Suddenly, we can't travel that way. We're cut off by the water. The area that I went up with the men and came back down, I can't travel that way any more. We were probably gone a couple hours. . . .

I had to find another way around. I'm traveling and I said, "Look, there's only one thing for us to do now." In the mines [they dig] what we call crosscuts or chambers from one heading to the next. . . . As you advance sixty feet with your parallel chambers, you have to make an opening to the next chamber for ventilation. Then you have to go back and seal the last one with cinder blocks and cement. They wall those crosscuts with cinder blocks for ventilation so they can carry their ventilation ahead with them all of the time. The air has to be up in the face of the chamber all of the time. . . .[4]

So what happened, we were in this heading and they're all sealed off. I know I have to go in that direction. So, we had a sledgehammer with us and I told Jimmy, "Start knocking those cinder blocks out and we'll see how high the water is." He was going to start in the middle. I said, "No, don't start in the middle, start up high, keep knocking the cinder blocks off." When he got about four feet from the bottom, the water started to flow over but it wasn't fast water, swift water, it was just water that was rising. I said, "Let's go through here. We have to get up in that direction."

That Jerome Stuccio he was pretty old, well into his sixties. Between trying to push him through the water—because by this time there was old timber floating around—and trying to carry those tools, trying to keep them out of the water. Oh, we had an awful time. Finally, as we were going, this chamber is going up a slight grade and the water seems to be dropping, getting lower and lower and lower. We weren't quite out of the water when I noticed a couple of lights ahead. I called the name of one of the fellows that was in our crew. It was a different voice. I knew it was the mine foreman's voice, Frank Handley. I said, "Who is it?" He said, "Frank." I said, "How did you get down here?" He said, "They opened up a hole over there [and] we came down through the air shaft."

We threw our tools down and we went up to him. He said, "Yeah, one of the guys you left there opened up a hole and he climbed up the side and he ran over the engine house and told us that there were men on the bottom so we came over." What happened, those men that I left there started pulling this debris down, rocks and trees and everything. From in the mines it looked as though it was filled right to the top. Instead, there was about twenty feet of fill in there. When they opened it up, they got up on top of the fill, but they still had another forty feet or so to go. This was an old shaft where trees were growing out of the side of the rocks. He [Amadeo Pancotti] took his boots off and by grabbing ahold of those trees he made his way up to the top. . . .

We met those other inspectors down there and I asked them about Myron Thomas and his crew. I said, "Did they get out?" They said, "No, we didn't see them." I said, "Well, there's only two things could happen. If they tried to get out through the Schooley down the Marcy Slope they wouldn't stand a chance because the water was going right down. Or, they're lost somewhere in the old workings. . . ." I said, "We broke up because I had to stay back with some of the older men, and Myron Thomas said he knew where the air shaft was so I figured I'd meet him at the air shaft." I said, "When I got there with my men there was nobody there."

So what happened, in order to get up to this old Eagle Air Shaft you had to go in so far in the main heading road and then go up one of the old chambers [and] make a turn to the right, a ninety-degree turn. Myron probably got excited, missed a turn, and just kept going straight and that's how he got lost. . . . He didn't realize he was in the old workings. When you're in the old workings everything is new to you [and] it's hard to try to determine which direction to go. They went to the end of the Ewen workings. They couldn't go any farther. They went all the way up almost to the new bridge in Pittston, the concrete bridge. That's a couple of miles. When you're traveling in the old workings you're zigging and zagging. Maybe you went five or six miles or you hit a caved area that you couldn't go through [and] you have to find your way around. It's not that you're going on a straight line. They finally realized that they were lost so they were making their way back.

## ORAL HISTORY: JOE STELLA

Joe Stella (figure 22) provides further insight into his march to the Eagle Air Shaft by revealing the presence of a mystifying white light that guided his way. (From Joe Stella, taped interview, March 15, 1999, NPOLHP)

JOE STELLA: As I was traveling, I'd have my map and there are what you call control points, these numbers you see on the map here. This number here would be on the roof in the mines and as I'm going by I'm checking myself to make sure that when I find a control point we follow it. I look at my map and say, "Okay, I'm right here. I'm going in the right direction." And as I'm looking up I could see what to me looked like a halo going right out. It started out real small and it went out there, really expanded, and it got large as though I could see daylight. It may have meant just keep going, that there [is] a place to get out, but I didn't realize anything at that time until later on, you know. I said, "That's why I didn't feel any pressure, nothing that day."
ROBERT WOLENSKY: It calmed you?
STELLA: Yes, as if I knew what to do. I didn't have to worry about should I go this way or should I go that way. I just went the one direct route right up there.
RW: You saw it at all times?
STELLA: No, no, no. Just. . . I don't know. I didn't see it all the time, but I couldn't even tell you how long I saw it, but I know I saw it.
WOLENSKY: Did you see it for five minutes, ten minutes, or every so often?
STELLA: Oh, every once in a while I'd see it. It seems to me every time

Fig. 22. Joe Stella holding the original photo of his rescue, 1999. (Courtesy of Stephen Lukasik)

I'd be looking up I could see a light out there.
WOLENSKY: So you went straight, followed that light?
STELLA: Well, I didn't follow the light, no, because I had to check myself with those control points on the map. Sometimes they wouldn't be there because maybe [in] that section there might have been a bad roof [and] eventually there's some rock that gets loose and can fall. So that's how I was checking myself to make sure that I was going in the

right direction.
WOLENSKY: But this light, do you think it guided you? Is that what you're saying? Or did it just give you confidence?
STELLA: I think it gave me confidence because I felt as though, just don't give up, keep going because there's a way out.
WOLENSKY: So it was more inspirational . . .
STELLA: Right.
WOLENSKY: Was it actually showing you which way to go?
STELLA: Right. Oh yeah, yeah. I never told anybody, you know, just a few people and that is just lately. In fact, I don't think maybe until about a year ago that I told my wife about it. I kept it to myself. I didn't think much of it, but then as the years have been going by and by [I decided to speak about it]. I don't know why [I should keep it to myself], I said, because I never had that happen to me before so it must have meant something at the time.

### NEWS STORY: AMADEO "PAUL" PANCOTTI

Paul Pancotti (figure 23) fled toward the Eagle Air Shaft as part of Joe Stella's group. He became one of the heroes of the rescue because of his nearly vertical, fifty-foot climb out of the air shaft. The following newspaper report drew upon an interview with Pancotti immediately following the disaster. (From "He Climbed Out, Got Rope to Aid Others; 'God Was With Me,' Says Exeter Miner in Recalling 50-Foot Climb," Pittston *Sunday Dispatch,* January 25, 1959)

"Rope, get some rope." And with that cry, which rang out Thursday afternoon high atop the hill overlooking the grim workings of the River Slope of the Knox Coal Company in Port Griffith, sprang hope that there was still life for some of the 45 men who were, at that time, entombed in the workings which were being deluged with water cascading through a break in the bed of the Susquehanna River over a chamber off to the left of the original foot of the slope.

It was a joyous cry for with it unfolded a tableau, which eventually led to the rescue of 33 of the trapped miners.

Amadeo Pancotti, a 50-year old company miner from 108 Schooley Street, Exeter launched the rescue activity when he miraculously made his way up a 50 foot abandoned air vent of the old Eagle workings.

Pancotti, who had clawed his way along with three fellow miners through 30 feet of debris to reach the foot of the air shaft, emerged at the foot of the huge bluff overlooking the Susquehanna River, about 30 yards north of the point where the ice-swollen river had crashed through its bed into the River Slope workings.

Bill Hastie, Knox Coal Company employee who was to have started work at River Slope at 2 o'clock on the ill-fated day, saw Pancotti emerge from the air vent as he waited to direct a Lehigh Valley train which was coming to the disaster scene.

When interviewed by the *Dispatch* photographer Saturday afternoon at his home, Pancotti recalled that he spotted Bob Groves' son-in-law (Hastie) upon emerging and informed him that three of his companions were at the foot of the air vent and needed a rope to get to safety.

Hastie started to yell for rope and the cry was repeated by an onlooker who was at the top of the hill and soon reached the office at the River Slope site where a rescue team with a cable was quickly dispatched to the scene.

The *Sunday Dispatch* reporter was on the top of the hill photographing the whirlpool action of the river breakthrough at the time the call for "rope, get some rope" went up and quickly headed for the site of the air vent which was to ultimately provide the escape route for 32 workers in addition to Pancotti.

Pancotti made his remarkable climb to freedom after locating the air vent along with six other companions: Joe Stella, a surveyor for the Pennsylvania Coal Company who happened by chance to be working at the Knox when the breakthrough occurred and played a major role in locating the air vent; John Elko, Joseph "Lefty" Soltis, Louis Randazza, Jerome Stuccio and James LaFratte.

Their dash for freedom began earlier in the afternoon when cry of "everybody out" interrupted their work.

The group, which originally numbered upwards of 30, headed toward the River Slope but rapidly rising water, which was soon swirling around their waist in some spots, brought the realization that the river had broken through.

Mine Foreman Myron Thomas of Taylor advised the group that they had better head for the Eagle Shaft and the incoming water cascading in "like Niagara Falls" as Mr. Pancotti put it, got them off to a fast start.

As they toiled their way to high ground, with icy water ever present, the group started to break up as some made better time than others. Eventually Pancotti found himself in "the group of seven" and Stella directed them towards the abandoned air vent, which had been filled with debris through the years.

Stella told them the air vent was 40 or 50 feet deep and when they finally found it, only a small funnel of daylight was visible through the debris.

While Stella and two companions headed back for tools to tackle the job of digging their way to the base of the air vent, Pancotti, Soltis, Elko and Randazza clawed their way towards the opening and

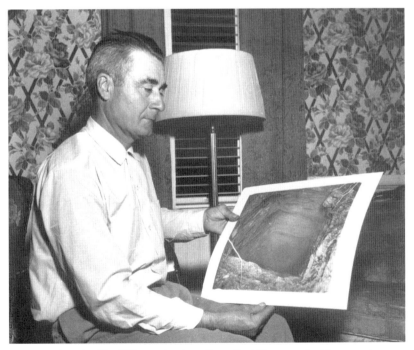

Fig. 23. Amadeo "Paul" Pancotti, miner, Knox Coal Company, looking at a photo of the Eagle Airshaft. (Courtesy of Stephen Lukasik.)

eventually made their way to the foot of the shaft.

They began yelling for help but soon realized their calls could not be heard so Pancotti, in sheer desperation, decided to make the treacherous climb to the top of the icy precipice.

Climbing upon the shoulders of Elko, Pancotti started the hand over hand ascent, which carried him upward to freedom.

Looking back on the dreaded day, Pancotti has only to say that "God was with me" and if you were to look down the air vent Pancotti climbed, you'd be quick to add "and he must have really did some pushing as you made your ascent."

Mr. Pancotti, who worked in the mines for the past 33 years, has been employed by the Knox Coal Company for the past 10 years and spent the past four years working at the River Slope operation.

### ORAL HISTORY: WILLIAM "BILL" HASTIE

Bill Hastie (figure 24), West Pittston, an off-shift Knox Coal Company laborer, was guarding the Lehigh Valley Railroad tracks around the Eagle Air Shaft when he met Amadeo Pancotti who had just ascended from the mine. (From William Hastie, taped interview, July 31, 1989, NPOLHP)

I was assigned to patrol the Lehigh Valley railroad tracks, which ran very close to the river in the bed of the old North Branch Canal. I had orders to stop all traffic, rail or foot, because [of the] hole in the river bottom right against the bank. It was eating away the bank and it well could have eaten away under the railroad tracks.

Sometime later. . . on the way upriver I encountered Amadeo [Paul] Pancotti who was coming down the tracks. . . . He was dressed in mining clothes and I thought he was a second shift miner waiting. . . off-shift, really, because of the disaster, and satisfying his curiosity. I stopped him and I said, "Paul, I can't let you through because the tracks may cave in." He blurted something out very quickly which I didn't understand. Paul spoke in broken English. I began to scold him a little and he interrupted me, and this time with great vehemence, got it through to me that he had escaped from the mine!

He had escaped by way of an air shaft to the old, long defunct,

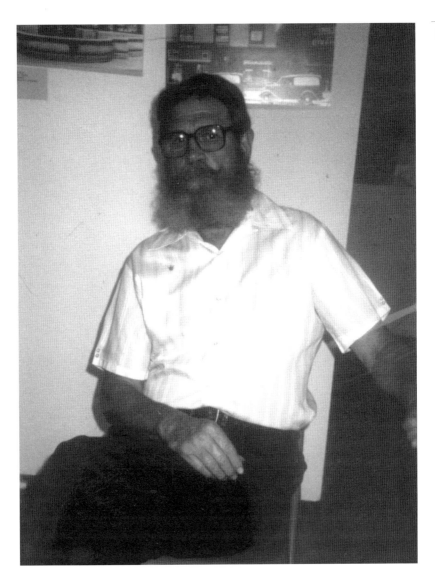

Fig. 24. William "Bill" Hastie, laborer, Knox Coal Company.

Eagle Shaft mine. . . . He had climbed fifty-two feet up the sheer rock face [using] finger and toe holds. Now, probably the worst part of his climb was trying to get over the top because he'd only have loose dirt to clutch at. And if I'd just gone over there on my first impulse—because I had been in that area earlier and saw steam coming from the air shaft—I'd probably [have] been there to help him up over the edge. Well, I've kicked my seat ever since for not having done that.

But at any rate, he told me that there were five [six] more men at the foot of the shaft. I immediately called up to some men that were along the top of that cliff, night shift mineworkers, and yelled as loud as I could to "Get rope! Get rope!" They ran off, scurried off in the direction of the engine house.

I hurried over to that hole and I soon had company. They got rope to us in a hurry. It wasn't exactly rope, it was heavy electric cable with a heavy rubber covering but it sufficed. It turned out there were only [three] men at the foot of the shaft then. We got them out. We lowered the rope and got them out.

### Notes for Chapter Two

1. Unless otherwise indicated, the photographs of oral history participants were taken by Robert P. Wolensky.

2. Inaccuracies in the oral history and other texts are rectified by inserting the correction within brackets immediately following the error.

3. See the interviews of Thomas Burns and Joseph Kopcza later in this chapter regarding problems associated with Handley's signal to lift the cage to the surface.

4. Stella has often stated in public addresses that, "This is the only time that we really thought that we were done for, because we couldn't find our way back to the Eagle Air Shaft."

# Chapter Three
## Delayed Escapes and Missing Men

*If you want to see twenty-five men on their knees thanking God for saving their lives, you should have been there.*
　　Myron Thomas, assistant foreman, Knox Coal Company

*I had dynamite. I told the guys—you ask my kid brother—I said, "I have dynamite." I said, "I am not drowning. If I have to put a wire in the battery, I will blow my head up before I go out and drown." Drowning is an awful death....*
　　John Gadomski, miner, Knox Coal Company

### News Story: Myron Thomas

　　Myron Thomas (figure 25), of Taylor, an assistant foreman for the Knox Coal Company, led twenty-five men to safety in a seven-hour trek to the Eagle Air Shaft. Shortly after his rescue, Thomas recounted the following story to reporter Ed E. Rogers at the Pittston Hospital. (From "Mine Disaster Hero Tells His Story: Thomas Talks to Newsman," *Scranton Times,* January 23, 1959)

We were working about 250 feet from the slope entrance when suddenly a motor runner and brakeman rushed in and told us to get everybody out of the mine right away. He didn't explain why.

I sent the motorman into the Pittston Vein to warn the men there while I hurried to the lower vein—the Marcy Vein—to warn my two crews (seven men) who were working there.

When I reached the Marcy I found the lights had gone out at the foot, but one crew was out and I saw the lights of the other men coming along the gangway.

I strained my tonsils yelling to them to come out. When they got to me, we all headed for the higher levels of the Pittston Vein. The noise of the water gushing in sounded like two fast freight trains passing in a tunnel.

We worked our way to the River Slope but found our way blocked by water. Then, we knew our only salvation, the only possible way out was the old Eagle Shaft. That was at a higher elevation.

When we got to the substation [a section of the mine] I found that seven of my men were missing, but I knew they were with the surveyor [Joe Stella] and he had maps of the whole place.[1]

For hours we walked over cave holes and through water up to our knees. To save power in the batteries of our headlamps, I ordered half of the men to put out their lamps. We used the lights in shifts. I had only part of a map, but I knew the general direction of the Eagle Shaft.

Some of the men became excited and I had to try to keep them in line. I knew we had to remain calm. Once I had the men sit down on a cave-in with their lights out while another man and I went scouting. To tell the truth, I didn't know where we were, but after an hour we found our way back to the main group.

One of my men had worked in the section when the Pennsylvania Coal Company operated there. He led us down an old slope to the Marcy Vein again, but the water was too deep and we had to head back to high ground.

About an hour before we were rescued, I went scouting alone. Suddenly, I noticed some chalk writing on a rotted door frame. It said "To Eagle Shaft." Then I knew we were on the right path.

I yelled to a man I had stationed about 100 feet back in the pas-

## Mine Disaster Hero Tells His Story

### Thomas Talks To Newsman

**By MYRON THOMAS**

(As told to Times Staff Writer Ed E. Rogers)

(Myron Thomas, 42, of 492 Union St., Taylor, the assistant mine foreman at the Knox Coal Co. operations in Port Griffith, led about 30 men to safety last night after the rampaging Susquehanna River broke through its banks and flooded the mine. This is his story as told to a Times reporter at Pittston Hospital).

We were working about 250 feet from the slope entrance when suddenly a motor runner and brakeman rushed in and told us to get everybody out of the mine right away. He didn't explain why.

I sent the motorman into the Pittston Vein to warn the men there while I hurried to the lower vein—the Marcey Vein—to warn my two crews (seven men) who were working there.

When I reached the Marcey I found the lights had gone out at the foot, but one crew was out and I saw the lights of

Fig. 25. Myron Thomas, assistant foreman, Knox Coal Company. (Courtesy of the *Scranton Times*)

sageway, "I found the way to the Eagle Shaft, bring the men up here."

Then we realized that the water pouring into the mine was pushing air toward the Eagle Shaft. We followed the air, crawling over debris and digging through cave-ins.

About 100 feet from the shaft opening, we spotted a light coming towards us. We yelled. It was a federal mine inspector and a search party. We were safe.

If you want to see 25 men on their knees thanking God for saving their lives, you should have been there.

### News Story: Myron Thomas

Myron Thomas recounts the saga of his escape in another newspaper account. (From "24 [25] Saved by Foreman, Say Surviving Miners; Taylor Man Led Group To Safety In 7 Hour Journey To Air Shaft," *Times-Leader Evening News,* January 23, 1959)

Myron Thomas, assistant mine foreman at Knox Coal Company, and a resident of 402 Union Street, Taylor, was credited with saving the lives of 24 [25] miners, besides himself, when he led them from underground chambers to an air shaft. It took seven hours for the men to reach the safety of the air shaft.

Thomas and his two crews were the last to be brought to safety, reaching the surface shortly after 7 o'clock last night. His crews were working in the Pittston and Marcy Veins of the coal company's workings. The mineworkers first became aware of river water flooding their workings at about 11:50 a.m. yesterday. Thomas said the onrushing water and blocks of ice "sounded much like two speeding freight trains passing each other."

The mine foreman was with one crew in the Pittston Vein, which is at a higher elevation than the Marcy Vein. [He went] through a connecting passageway and ordered the crew working in the lower level to vacate that area and join the men in the Pittston Vein.

### Thomas Had Maps

From this time on, Thomas, who had several small maps on his person of the underground workings, led the men from one passageway to another seeking higher elevations. The veteran miner of 23 years experience, although only 43 years of age, followed the flow of air coming into the mine. He knew that by doing so he would reach an opening.

A number of times during the seven hours of their ordeal, the men were uncertain of their exact location in the maze of passageways through which they attempted to reach the air shaft.

Thomas ordered his men to rest every 20 minutes or so and to turn off their electric lamps. During these rest periods, the miners discussed among themselves their next move. Thomas also conserved a supply of batteries by having only every other man use his lamp at a given time.

In an interview with a reporter, Thomas stated that the swirling river water with terrific force "pushed my men all around the place. We grabbed on to old electric power wires which had long since been out of use to keep ourselves from being thrown off our feet."

### Survived Other Disasters

Shortly after World War II, Thomas was in another mine disaster when 10 men were killed in an explosion at the Knox works. The mine foreman was in a different vein on the west side of the Susquehanna River at the time of the explosion but still was tossed into the air by the heavy blast. However, he was not injured. The only survivor in last night's mine disaster who actually was injured was John Pientka, age 51, of 377 North Main Street, Plains Township. He was one of the miners who escaped with Thomas.

When the onrushing of water was noticed shortly before noon yesterday, Pientka grabbed hold of a wooden ladder to prevent himself from being washed away in the flow of water. He said he first thought he heard cars rolling down the incline, but instead saw huge chunks of ice in the turbulent water rushing toward him.

As he grabbed the ladder, said Pientka, and began climbing to a higher elevation. He said he slipped and twisted his ankle. He also received abrasions to his left leg and hip. Pittston Hospital dispatches listed Pientka as being in good condition today. X-rays are to be taken to determine if Pientka suffered any fractures. Asked if he were suffering pain from his injuries, Pientka said, "I don't care if both my legs were broken. I'm just glad we got out. Only God could have helped us."

Another survivor who reached safety with assistant mine foreman Thomas was Merle Ramage, 214 Elizabeth Street, Pittston. The father of four children, Ramage is well known in the Pittston area, having played football for Pittston High School in 1937 as quarterback.

### Hope Rests With Burns

He also credited Thomas with leading the two crews to safety. Ramage mentioned the name of Frank Burns of 19 Elizabeth Street, Pittston, as being an "old-timer" in the Knox operation. Burns actually is an inspector with the Pennsylvania Coal Company and was checking pumps in the Knox operation yesterday.

Ramage said that Burns knows every inch of the subterranean area and can find his way around the workings without the aid of maps. It was Ramage's contention that if anyone could bring out the [lost] men, Burns is the one.

## Oral History: Joseph "Joe" Francik

Joe Francik (figure 26), a laborer from Pittston, recalls his flight from the mine as a member of Myron Thomas' group and being taken to the Pittston Hospital into the arms of his anxiously awaiting family. He indicates that Joe Stella's group walked behind Myron Thomas' crew. (From Joseph Francik, taped interview, June 26, 1990, NPOLHP)

JOE FRANCIK: Well, the day of the disaster we were working down in our chamber. It was about eleven o'clock. We loaded our coal and we started working on the timbers. You'd had to put your timber up to protect yourself so the roof didn't fall in on you. Around that time a man comes down and walked by our place and hollered in, "Everybody out!" So, well, everybody out. Usually when that happens, when they come down like that [and] they say everybody out, we think [that there is a problem with] the big air vent that supplies air into the mines. Ventilation.

So we took our time and put our tools away. Each miner, night shift and day shift, had their own tool box. So, we put our tools away. [We were] not running and not scared or nothing. . . . Of course, we didn't know that the river broke in. So, we went out and got on the heading road and we walked a-ways. On our way to the May Shaft we met a couple of men including Myron Thomas. We all huddled together and they said we had to get out. Not [a word] about the river coming in.

So we all started walking out toward the May Shaft, the way we came down, off the cage we called it. As we start walking, jeez, I looked up the road [with] our light and you could see the water coming down! At first it was a little low, but as we proceeded, the whole group of us, it got pretty high—right up to our chest and big icebergs were coming. We definitely knew the river broke in. . . . The boss said we can't get out the May Shaft, we have to try to go for the Knox [River] Slope.

We [walked up] an incline, a side pitch they called it, walked along that, crawled along there. The coal was mined out. There were props there and everything else. We didn't walk too far. Well if you ever were to Niagara Falls it's misty and noisy, but boy when we got

Fig. 26. Joseph Francik, laborer, Knox Coal Company.

up there all we could see was mist and thunder, and water was coming in. And the ice! We couldn't see [very far]. We couldn't go that way because we heard the noise. That's where the river broke in. We had to back track. It was about a mile and a half from where we were working. . . .

We were all in the group. Joe Stella had a map and he remembered the Eagle Shaft. So, we all grouped together to head for the Eagle Shaft. Of course, us miners wouldn't know where it was. Joe was with us—which is lucky—but Myron Thomas remembered a little bit about the shaft too. We all started for Eagle Shaft. In the meantime, we got separated. Joe Stella's group had seven men. We got lost. They were behind us and we made a turn instead of going straight ahead. We made it seemed to me like a left hand turn. Joe went the right way. He landed there at the Eagle Shaft.

We were following Myron Thomas. He figured he knew where to go, but we couldn't see their [Stella's group] lights or nothing. Joe did have a couple of older fellas with him. Everybody was kind of in high gear, you know, the water was coming in and we figured if we got to the Eagle Shaft, we'd have a chance to get out. We made the left hand turn and got lost. We wandered around there for seven hours in through the chambers, all around. Lost. We couldn't find our way out. . . .

A fella can sit here now and smile and think, but it was no fun. After a while we just kept on wandering and taking all kinds of avenues where we saw a hole in the chambers. Maybe it's this way or that way. We couldn't find it. We just sat down and rested for a while, then you had your own thoughts. You know what them thoughts would be—about the family and what you do to get along, and will they be okay and so forth. And you just [thought], that was it—you figured that you're done. . . .

We didn't stop much because we had to keep on moving. We kept on walking around, probably in circles, because you'd go in one chamber, out the [other]. The mines had chambers all mined-out to just pillars. You'd walk on the one side. You'd come out, there would be a block of coal. Then about eight feet there would be another block. You'd be just walking and you could walk for miles before you knew where you were at. We walked quite a number of miles. We didn't have any food. . . .

ROBERT WOLENSKY: How deep was the water?

FRANCIK: I'd say up to my chest. But that was going up to the slope towards the May Shaft. [It was] about fifteen to twenty minutes before we came up to higher ground, the next vein you could climb up out of the water. Up into the Big [Pittston] Vein.

WOLENSKY: Did you ever have to swim through the water?

FRANCIK: No, no. No, we just walked. Held on the sides, just went in line and we got out of there.

WOLENSKY: Was there a current?

FRANCIK: You had to watch the icebergs coming, push them on the sides. You'd hang on the side of the road. There's timbers there and you'd just grab [on]. We weren't too long in it. Then we got up higher, the next vein, we got up higher ground. . . .

So, in the meantime, [we were] just sitting there and Myron Thomas said, "I'm going to take a walk and look around once more." He went out a-ways and in a few minutes he hollered back. He said, "I found a piece a wood here." And written on it was Eagle Shaft. It was pretty well aged. Well, we heard that and all of our spirits picked up. So, we all gathered and followed him.

You could see as we were following him that we were kind of going up grade. We were figuring, well the water is coming behind us, but it's downhill, you know? We were not cornered in a hole or anything. We were going up. When we got in a little further we heard fellas hollering and [we saw] lamps. Well, that was where Joe and the seven [six] fellas got out.

You were pretty shook up. When you got up there [outside] you'd see all the people, all the workers and people there. Up on the hill you'd see crowds of people that were looking down, watching to see everybody coming out. It was a good feeling once you got out of that hole.

WOLENSKY: Was your wife there?

FRANCIK: Oh, yeah, they were up on top of the hill. Oh, the way she told me it was a freezing day. About ten above zero and windy. But all the people stood there and just waited. She was there with my two children. . . . They rushed us in the ambulance up to the Pittston Hospital. First, shock. The doctor had us stay over night. They gave us some medicine and well, first shock, you know, we rested. Then I saw her [wife]. They [family] came in. It was good to see them, too.

WOLENSKY: When did she know you were one of the trapped men?

FRANCIK: She was home and she heard [it] on the radio. The way she told me, it was terrible. She was waiting at least seven hours.

### Oral History: George "Bucky" Mazur

Bucky Mazur (figure 27), the youngest person employed in the mine, hired on as a laborer shortly before the disaster. One of four brothers employed by the Knox Coal Company, he and his half-brother, John Gadomski, were part of Myron Thomas' group. (From George Mazur, taped interview, December 20, 1988, NPOLHP)

GEORGE "BUCKY" MAZUR: I was laid off at the time. There were no jobs available so I went down to the mines until I was going to hit twenty-five years of age, which was February 22, 1959. I went in the mines about the second or third week in December 1958. I worked there two days and got hit with a shaker chute and broke my leg. I didn't work [again] until after Christmas.

I went back a day earlier than I should have. Now my brothers worked there, they were miners all of their lives—my brother Eddie, a brother Danny, and brother John Gadomski—same mother, different father. I didn't know enough to be really scared when the things were going on down there.

I was about twenty-four years old. My brother Eddie was on the afternoon shift and the other brother Danny was supposed to be working that day. But what happened, a friend of his owned Quality Bev-

Fig. 27. George "Bucky" Mazur, laborer, Knox Coal Company.

erage in Wyoming, and they were flooded and he was working there all day and night helping to pump the water out. He skipped that day or he would have been trapped down with us, the three brothers! I was working with my other brother John at the time. There would have been three of us instead of two, and the other one was coming in for the afternoon shift.

I was just the labor force [a laborer]. I was only supposed to be down there about two or three months. My brothers said, "See what it's like." They said, "Maybe it'll be good experience, go down and see what it's like so when you get this [other] job you'll appreciate it instead of going down in the mines."

I got up late in the morning [on January 22] and drove over to the shaft. I was a little late so my brother told the foreman, "[When he arrives], tell him to carry some dynamite down into where we're working in the lower Marcy." I changed and picked up the dynamite, went down in the cage into the lower Marcy, and walked down to where we were working. At the time they were robbing pillars and doing different things. Well, we were working in the chambers and they came down and said to get out.

Now if you know the Wyoming Valley, we were working on the other side of the river, on the west side, which is Wyoming, Exeter, around near Saint Cecelia's Catholic Church, in between the mountains of the back road and Wyoming Ave. If we knew what happened, we could have gotten out the Schooley Shaft maybe about three-quarters of a mile or a mile away but they didn't tell us what happened, they told us to get out quick. We got on the motor when the guy came to tell us to get out and we went to a certain point and he had to go someplace else, so we started walking. It was like walking along the edge of the river and walking underneath the river to get back onto the other [east] side. When we got under the river to start coming up, the water was coming down, gushing right down the main chamber so we had to go to higher ground. . . .

When we walked under the river we started to hit the slope coming up. That's when in fact we would have gotten killed if the wires [were live]. We had to grab the cables or we would have gotten washed down into the lower Marcy. . . . It was a good thing we did because they [some victims] went to the lower Marcy. . . .

Nobody knew there was a break. In other words they told us to get out, they didn't tell us water is coming in. They just said something's happening, get out. So when we walked under the river we got hit with water and that was it. . . .

[We connected with Myron Thomas and Joe Stella.] We worked our way into the chambers and couldn't find the Eagle Air Shaft. Joe Stella and a few of them [did] but that group broke away somehow and they had the map and we never had the map. They knew where they were going and they found the air shaft. . . .

We were trying to find the air hole, going all day. I'm talking [about] walking [and] walking—what was it, a good six, seven hours? Every place we went I was up to my chest [in water]. Some of them were up to their neck in the water and trying to wade through it. There were chunks of ice big as desks that were hitting you, but we could never find the right place. We were always walking to the higher elevation to try to keep away from the water. . . .

They wanted to try to find a high spot then maybe they would come down and try to save us from there, dig a hole down or a shaft or something that they would get it, but it never happened. What happened, we worked our way around and we were just walking from noon until about six or seven o'clock at night into the higher veins and to the lower and upper Marcy we finally found the air shaft and that's how we got out. . . .

The only time that I got scared was when we found where to get out, and there was nothing but water there. It was fresh water [not mine water] and I remember being a little bit scared because I figured, "Hey, we found where we're going; now there's no place to go." But then they walked up a hill where there was water coming down [and] there was the rock to [stand on to] get out. . . .

There's only about two or three [men] that really should have got killed in the whole thing after the whole mess was straightened out. They were engineers; Burns [was one.] They shouldn't have been there that day but they had a meeting or something [and] they said, "While we're here we'll go down and check the [pump] motors." So they went down to the lower part. When the river came in it sought its lowest spot and they had no way to get out because of where they were. But [the others] should have never got killed. A lot of them that got killed went back for their tools. They went back and they got trapped. They never got out. . . .

ROBERT WOLENSKY: Were you at all angry with the company for putting you in that kind of jeopardy?

MAZUR: No, not really. The only thing I was angry with was the group that had the map and split, that's the only thing we were angry about really at that time.

WOLENSKY: You thought you could have stayed together?

MAZUR: Right, whatever happened . . .

WOLENSKY: Why did you split up?

MAZUR: I have no idea. All of a sudden somebody went one way and somebody went another way.

WOLENSKY: Was it a kind of pandemonium when you're making these decisions?

MAZUR: Not really, I was following Mryon Thomas. Stella was with [the first] group which means he had the map, maybe the other ones got it first, maybe Mryon Thomas, either one of them, I just . . . Time erases . . . .

WOLENSKY: Did you see any rats as you were making this excursion?

MAZUR: This is the only thing that shook me up. My brother took about three or four sticks of dynamite and caps and the plunger. He kept [them] with him. Then he said, "If we can't get out, we're not gonna be eaten up by rats." I said, "You gonna blow us up?" He said, "If we don't get out. You don't want to get eaten up by rats?" In other words, when we go to the high points and the water is coming, the rats are going to the high points, and if it gets to the end, you're not gonna drown. Just blow yourself up. . . .

WOLENSKY: So he was prepared to do that?

MAZUR: Yes he was, yes he was.

## Oral History: John "Stover" Gadomski

John "Stover" Gadomski (figure 28), a miner from Wyoming, had been working for the Knox Coal Company for nearly four years when the disaster occurred. His crew included Stanley Roman, John Gustitus, and his brother, George Mazur. All survived as part of Myron Thomas' party. Gadomski criticizes the separation of the Thomas and Stella groups. Mrs.

Gadomski briefly joins the interview toward the end. (From John Gadomski, taped interview, December 22, 1988, NPOLHP)

> JOHN "STOVER" GADOMSKI: I didn't want to go to work that day. I picked Stanley Roman up at his house because he lived right here. We worked together and we met my other worker, Yosh [John Gustitus], over there. Bucky, my kid brother George Mazur, was supposed to come in later.
>
> The river was high as we were going over the Wyoming Bridge and I said, "Stanley, you see that river?" He said, "Stover [nickname], I don't know; let's go back to the bar room." This was about six o'clock in the morning. At that time the bars around here were open at five o'clock. Then Stanley said, "Let's try to make it in." I said, "Okay."
>
> My kid brother Bucky didn't come until [later] . . . We went in [and] started robbing one pillar back. We had nine cars loaded I think, it was about 11:30 a.m., and we had three more to go. We had enough along the shaker chute to finish the shift. Twelve cars we needed. I heard motorman coming in. I said, "What the hell, is he changing on us already? He just brought us a load of empties." Lefty came in—I don't know his last name [Soltis]. He said, "John, everybody out." I said, "Why?" He said, "I don't know but everybody out." So I give the guys a shake on the shaker. I was out there topping that time because we usually turned around [took turns]. Stanley was a miner, John Gustitus was a miner, and I'm a miner so you have a miner in the chamber at all times.[2]
>
> So I said, "Yosh, something's the matter." I noticed there is no air or nothing. He said, "Yeah, it's awful still isn't it?" We were walking out and he said, "I hear noises or something. What the hell is that?" We start walking and then walking faster and faster. So we get by the Marcy Slope. We were down a slope from the Pittston. "Holy Christ," I said. "Yosh, you see what's coming down the slope?" Icebergs, twelve inches thick and maybe six to eight feet wide, and long. Holy Christ! We start jumping around them [and] we got to the top. They weren't moving fast in the water. Some were getting pushed by the others and they would be sliding down one on top of the other. Jeez!
>
> Myron Thomas was on top [in the Pittston Vein] with a bunch of guys—Stella, Pancotti and a bunch of others. I said, "What the hell

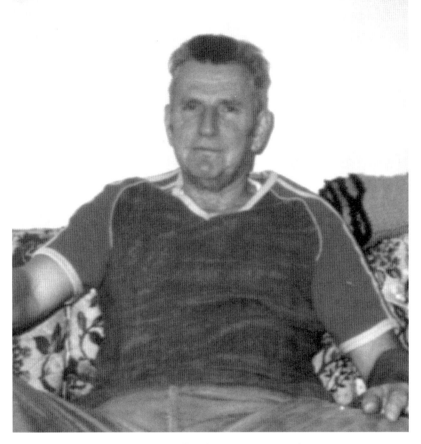

Fig. 28. John "Stover" Gadomski, miner, Knox Coal Company.

happened?" He [Myron] said, "John, the river came in." I said, "Oh boy, goodbye! Where's the nearest exit?" He said, "We ain't got it; we could try the River Slope." So we went that way. He said maybe we could bypass the water and go up through the crosscuts and bust a wall and get out. The water came in! The first crosscut down and you should have seen it rolling! Holy Christ! Oh, I can't describe it. Like Niagara Falls, just a complete roar. And stink? It stunk terrible!

So we got out of there quick. We went back. Then Stella started looking at the maps. He had the maps on him. . . . Stella said, "There is an Eagle Air Shaft." Myron said, "Yes, I was up that way a couple times." "Well," I said, "Let's head for it." I said, "We got twenty-six [thirty-three] guys; we could dig our way out. It is only eighty [fifty] feet. What the hell? In shifts we will dig our way out."

So Myron said, "Come on, we'll go back to a couple chambers [and] we'll get the picks, we'll get some dynamite, we'll get some props." Everybody carried something that we needed in case it was caved, right? Well, we were going for the stuff [and] we left Stella, Jerome [Stuccio], Lefty Soltis, Pancotti, and two other guys, I don't know their names. Six [seven] of them were supposed to wait for us. Right here. Don't move until we come back. We came back, they were gone!

I don't know where they went. They said they thought we got drowned and they left us there and they went on their own. They knew where the Eagle Shaft was. Stella was the big boss, he was an engineer, he had the brains, he knew the spads [numbered metal markers on the roof of the mine]. All he had to do was look for the spads [and] he would get up [to the Eagle Shaft], right?

Myron said, "Where the hell did they go?" I said, "We ain't finished yet." I'm a trapper and I follow tracks. This is how we started, by following their tracks. Because nobody else was in there [so] the dust just lays like snow, but it's black and you could figure where those guys went. So we start following their tracks for a while. . . .

We were walking [from] maybe twelve o'clock until about seven o'clock. We were walking around water this way trying to get different ways. We were walking through crosscuts and I said, "What the hell are we doing here?" What the hell, we were walking around for how many hours? We didn't know where we were. I can't explain walking around with a bunch of guys smoking and bitching. You know you're not supposed to smoke in the mines [but] everybody smoked. I didn't smoke but there were guys on my shift [who] smoked. They pulled out cigarettes [and said], "You're going to die so you might as well blow yourself to hell [rather] than drown to hell." I mean, you're scared to death. . . .

Myron, Myron, Myron. I tell you, he was so shook up. He was shook up when he was in there. He was very nervous, he was. That poor guy took a lot on his shoulders getting these guys out. But he brought us all [out]. . . .

They didn't tell us the river came in. [If they did], the [Schooley]

Shaft near where we were working was five to ten minutes walk. I would have been right out. I checked that out a lot of times for safety's sake. We used to go over there and the guys used to talk to each other. They'd say, "If anything happens come out this way, right down here." We were working maybe four hundred to five hundred yards from that shaft. They didn't tell us why we should get out, where we should get out. So we went across the river.

[After hours of walking] my buddy said, "Hey, it's getting windy in here so we must be next to an opening or some goddamn thing." The wind was blowing like hell. He looked down [at] a door and there was an old chamber going up and there was water coming down. The door had "Egl" on it and it had an arrow. So I said, "Yosh, you know what that means?" He said, "Yeah, that's where the Eagle Air Shaft is." Myron said, "You are right." Sure enough, we ran into their [Stella's group] footprints under there. We went under a cave and we start getting pretty close. Then I saw two big lights coming, bright lights. I said, "Here comes the rescue crew." Two mine inspectors came and they said, "Oh, were we looking for you!" I said, "Brother, were we looking for you!!"

I said, "Where's them other six [seven] guys?" "Oh, they come out about 1:30 or 2:00 p.m." I said, "How could they get out so quick?" We were walking around down there like nuts and they were out three or four hours before us? We were out at eight o'clock. [I thought to myself], "That guy must be nuts." I said [to him], "They just left us about three or four hours ago." He said, "It may have been a little later but they're gone, they're out. They said you were lost." I said, "Who told you we were lost? They left us in there. They left us in there, that's it!"

Since that day every time I saw Stella or Pancotti, I said, "You left us in there." [They would reply], "No no, no. We were scared. We heard noises." [I said], "Oh, come on. You don't leave twenty-six guys. Come out with us. . . ." I never forgave him [Pancotti]. I never forgave Stella. I never forgave none of them.

MRS. GADOMSKI: You got to forgive, you're Catholic!

GADOMSKI: I forgive them, but I'll never forget it.

MRS. GADOMSKI: You can't forget it.

GADOMSKI: I'll never forget it. . . .

ROBERT WOLENSKY: So you would see Pancotti sometimes?

GADOMSKI: Oh yeah. Oh God, me and him, we were deadly enemies. I started a lot of troubles up the Club. When you're half loaded, I'd tell them, "You guys, you bastards." The Lithuanian Club [is] where we used to hold our [Knox anniversary] parties. "No, we didn't leave you." They wouldn't come right out and tell us they left us. "We thought you were gone someplace else, you were lost." Yeah, we [were] lost, sure we were lost; after you left we got lost. Oh yeah. . . .

They won't tell you [about it]. These guys are keeping it [a secret]. These guys, you know, they figure, well they don't want to get anybody in trouble. You can't get anybody in trouble anymore, it's all over. . . . No, they don't want to talk about it, but I'll talk about it. Oh yeah. Because I'm telling you the truth. Why should I lie? I have nothing to gain by lying. I quit the mines and from that day I never went back in the mines no more, that was it. . . .

WOLENSKY: Was there a time, John, when you thought you weren't going to make it out?

GADOMSKI: Yes. I had dynamite. I told the guys—you ask my kid brother—I said, "I have dynamite." I said, "I am not drowning. If I have to put a wire in the battery, I will blow my head up before I go out and drown." Drowning is an awful death. . . .

At about 7:45 p.m. I got out. It was dark. It was cold—must have been [because] all the wet clothes just froze to me. I come out of the mines, I grabbed my helmet and threw it on the goddamn ground and I said, "No more mines. That's it." I never went down a mine since.

## Oral History: Stanley Roman

Laborer Stanley Roman (figure 29) of Exeter recounts the tortuous trek to the Eagle Air Shaft with Myron Thomas. He presents a non-controversial account of the Stella-Thomas separation. (From Stanley Roman, taped interview, December 22, 1988, NPOLHP)

STANLEY ROMAN: I put in fourteen years all together [in the mines]. I started in Exeter at the Payne Colliery. From there when they closed down I went to Bernardi Coal Company. Very tough conditions there—more or less a "dog hole." And from there we went to the

Knox—that was in January of 1958, a year before the accident. We knew some of the men there and [that's how we got the job]. It was a good job; we enjoyed it until what happened. . . .

We mostly had good working conditions at Knox. I worked in there [fatal area] a few times and I told one of the bosses, "If you have any chance to send me there again, don't." If there was a shortage of employees at another section, they would fill you in. The pay was good but that was an awful wet place. If I knew then [what I know now], I'd have never gone in that one day. I worked in that area and a couple months later the river came in. . . .

This fellow came on the motor to warn us. I'm telling you if it

Fig. 29. Stanley Roman, laborer, Knox Coal Company.

wasn't for him we would have been stuck down there. He said, "Fellows, you gotta get out." We figured, "What the hell? He must have been kidding," you know. But we grabbed our coats, put them on and we had a little slope, it's maybe a five minute walk to get up there, and then we hit the main road where they take the coal out to the front of the shaft. That was when we said, "Oh boy, it must have come in," because we were [stepping in water]. It wasn't forceful. We were able to get out up that ladder. Just looking back, that's all you could see is a wall of water coming in. . . . It was all that debris and everything. That water probably shoved enough [debris] to block it and that gave us a chance to get out or otherwise we'd be still in there. I give that motorman the credit, though. . . .

We were working in the Marcy Vein. We were about the last ones to know about the river break-in and once we left we had to go up a slope and that's when we hit the water. It was about chest high. If we had known that the river had come in, within five minutes we could have been out the Schooley opening, but it was too late to turn back and go down there. We could have been out there instead of spending the whole day in the mines wandering around. . . . Bob Groves, I guess his heart sank when he didn't get news that we got out that way. . . .

We were led by Myron Thomas. We were way up on a pitch and we could hear the roar of the water down below us. In other words, we were gonna try to get out that way. Unfortunately, we didn't and we were getting to higher elevations as much as we could. Myron Thomas led us through to the Pittston Vein. We were able to backtrack and go down. In the meantime the water must have blocked some place and it was only up to the ankles. We went through the Pittston Vein and he was talking about the Eagle Shaft, so we were more or less circling the area. How many times we missed it we don't know.

That was more or less it until John Gustitus—he was one of the workers with me and Stover [John Gadomski]—busted through a wall and we felt the breeze. . . . See, when they blocked off the walls, they would put in cinder blocks and they have a wooden frame for an opening. . . . It's a small trap door, not a big opening. There were markings on the door, I couldn't recall what it was. That was closed, but it was no problem once he hit it. You could have opened it as far as that goes. That was when we hit the air. We felt a breeze and we went up the slope and we could feel that fresh air coming down. We knew that we were getting out some place.

[We then went] up a little ways [and] there was a cave-in and it was just narrow enough for the men to squeeze through. I remember that because we had to be on our belly to pull our way out. We crawled through it and some of the guys were a little fat, but they made it too. As far as I'm concerned, that was the end of it right there, I mean as far as getting out. Then it was clear sailing. It was just a matter of time when one at a time we started climbing [out]. We were the last ones to get out.

See what we did, we split up. We took the word of the man that was familiar with that area, Joe Stella. He was the foreman or something. He worked in that section and he had a blueprint of the area. Stella went with Mr. Stuccio and some of the older ones. We decided to go with the man that worked in that area, Myron Thomas. We had some men that worked there years ago, but they couldn't remember the section. . . . Joe Stella had the maps. Myron Thomas didn't have them, but Myron was familiar with that area. He worked in there a few times. Joe Stella said this is the way we should go.

We were picking up boards and whatever we could find to make a ladder if we had to. They were more or less arguing over which way we should go. We went with Myron because he worked that area. We figured there was a better chance, and, after all, we got out. That's all I could say. So I give both of them credit as far as that goes. . . .

It was pretty rough because it had caved in [certain] areas. This was just about as dangerous as the water coming in. Most of the men took it in stride. Our lights were going out so we were taking turns keeping our helmet lights on. Some of them would keep them off, like I'd have mine off and a couple others [would too] and we'd just follow the leader, whoever was ahead of us. We walked in single file. It was in a close group. Nobody was going astray or anything. We'd stop a little while to get our bearings to see where we were. A couple of these men said it was so long, but, all in all, the whole group was calm. There was no panic or anything. . . .

We were down there, with all this wandering around, from the morning until about 7:30 or 8:30 p.m. We didn't care about eating.

And anybody that had cigarettes was smoking them because they figured it was their [last] chance. That's a no-no, you know. But they all did that once in a while down there. . . . I said my prayers every time we stopped. I said them silently but, as far as that goes, I consider myself religious. That's the first thing I wanted to do when I got out. I wanted to go to church but my wife said, "No you better stay home," because I had a beating. I got in the tub and I stayed in for about two hours trying to get all that dirt off me. I was cold and wet. . . .

They took us to the Pittston Hospital [after we got out]. They had all the guys together. They were feeding them, some of them were fed booze but all I was getting was hot coffee. I remember that good. In the room I was in there were men that were laid up with asthma but they didn't work. They had worked in different mines. The doctors wanted to keep me in overnight and I said, "No way, I want to get home." So, what time did I get home there, about 9:00 or 9:30 p.m.?

Some of the men that I knew didn't go to work that day, which was fortunate for them. This one guy, he could have gotten out but he took his time to change his clothes and he never made it. He was an older man. Then Frank Burns, he was making his rounds and there was no chance for him. . . .

ROBERT WOLENSKY: Was there ever a moment when you thought you might not get out?

ROMAN: There was one spot where they had the pumping station. It was a big area and the pumps were there and at that time they were at a standstill. This was the area we were walking through water almost up to our waist. It was a dead end that wasn't being worked. I figured [this might be the end]. [Then later], you could feel the mist. When we hit the high ground you could feel that mist coming through from the river. Ice cold water. Instead of being stagnant it was starting to get fresh. So it was worrisome there for a while. But I think we were safe [towards the end] as far as that goes, because the Eagle Shaft was on real high ground. . . .

## Oral History: Edward "Ed" Boroski

Ed Boroski (figure 30), Dalton, retreated from the mine with Myron

Thomas. Although he suffered long-lasting leg damage from the ordeal, he recounts the story with a certain lightheartedness. He recalls that Joe Stella led the way to the Eagle Air Shaft, rather than following Thomas and his crew. (From Edward Boroski, taped interview, August 4, 1989, NPOLHP)

> EDWARD "ED" BOROSKI: A guy came running up the gangway. "The river broke at the River Slope," so everybody started to panic and run. I was talking to a guy named Sam Altieri; he was an electrician. I said, "Sam, Come on." "No," he said, "Wait until I get my tools, my tools." I looked around, he was gone. Where I don't know [but] he never came out alive. . . .
>
> We were working in what they called the Marcy Vein. . . . We wanted to go towards the [May] Shaft. A guy said you can't. He said it's flooded, the cage won't go over the top. So we backtracked, me and Myron Thomas and twenty-four other guys. In the meantime, surveyor Joe Stella and my own miner, Paul Pancotti, they looked for the Eagle Shaft. They escaped. We were in the mines about ten hours. . . . At first I was in Joe Stella's and my miner's group. Then, well, then I saw everybody come. Joe Stella went one way, we went the other way. I followed the greatest number.
>
> ROBERT WOLENSKY: Was there some debate about which way to go?
>
> BOROSKI: No, we had no way to go! We had to walk on the gob [waste piles]. The water was right up to the top. We had no way to go but up. If we went down there was no way in heaven . . .
>
> There was one time we had trouble. When we were trapped, we heard something like box cars or ice or something coming down. We didn't know what it was. But my God the wind and blocks of ice! We were afraid to leave the place, which was the only safe place around. Myron Thomas sat down and started to pray. I said [jokingly], "Look-it Myron, I want to see a television program called "Twenty-six Mule Team Boraxo" [*Death Valley Days*]. Come on, find the Eagle Shaft!" I figured that ought to do it! We wandered higher from elevation to elevation. But then on the last elevation, somebody spotted a sign knocked down on the ground. We picked it up and it said Eagle Shaft. . . .
>
> WOLENSKY: How did Joe Stella's and Myron Thomas' groups get separated?

Fig. 30. Edward "Ed" Boroski, laborer, Knox Coal Company.

BOROSKI: Well, Joe Stella was the surveyor. He knew the place. I left Joe Stella and Paul Pancotti for Myron and the twenty-five other guys. Why I left them I don't know. They were always before us.
WOLENSKY: They were ahead of you?
BOROSKI: Yeah, they were.
WOLENSKY: You stayed back with Myron Thomas?
BOROSKI: Yeah. When we were going we had battery lamps. Every fourth guy turned his light on so there was enough light. So that's how we scrambled in the dark. It was dark in the mine [but] we always had light. . . . One guy was carrying a box of dynamite. Fifty sticks! If we couldn't get out, there was no alternative. . . .
WOLENSKY: Were there any rats in the mine?
BOROSKI: Rats? Yeah. See the old saying was, "You go like the rats."

We knew they were going up that's why we kept going up. You see, when I worked in the mines my father said, "When you see a rat and you kill him that's a cardinal sin. Never kill him, he might turn out to be your best friend." And sure enough he was right.

WOLENSKY: Did you follow the rats a long way?

BOROSKI: We couldn't follow them far because they'd go underneath a rock. Some got out. They were all headed up so we started heading up.... We were waiting [resting] and in the meantime the water kept getting higher. We sat near the Eagle Shaft. I happened to glance. I saw a light shining in the water. I talked to myself, "No, maybe it's [my] imagination." I looked again; it was shining. It was Joe Stella. Thank God for a man like him. So I hollered to him....

One by one [we came to the surface]. I had ahold of the rope first. They were so anxious they pushed me on the side. I saw it [the rope] first. They pushed me on the side in the anxiety to get up. We were only [in] knee-high water. Outside were state troopers, four ambulances, and everybody. They took us to the Pittston Hospital.

They started stripping everybody because we were all soaking wet. I had a new pair of pants and a new pair of boots on. I only had them [the boots] two days. They had to cut them off me. I was all right. When I got to the hospital that was the first time I ever had a shot of whiskey. They poured it down my throat! We were in the hospital at eight o'clock. Why? Because the nurse took my pulse [and] I saw her watch. I thought, "I missed my show!" They wanted us to stay there overnight but I had a young family home. I wanted go home....

After the Knox mine, I got a couple jobs outside [surface work]. The doctor said, "He's like a drunken man." I said, "What do you mean?" He said, "When a guy goes on a drunk, the next day he has to have another one. You go back down the mines and try again!" I worked in two mines after that. . . . . . But when I was in the Knox [disaster] my legs were frozen. When I went back in the mines I found it hard walking so I got a job at Topps [bubble gum factory]. I worked there for fifteen and a half years until I retired....

WOLENSKY: What did you learn from the Knox disaster?

BOROSKI: Well, never give up hope, and never kill a rat!

# News Story: Eagle Air Shaft Rescue

The following report (figure 31) describes the scene on the surface near the Eagle Air Shaft. (From "Miners Found in Waist-Deep Water; Knox Men Are Discovered Near Air Shaft—Scenes At Hospital," *Times-Leader Evening News,* January 23, 1959)

People just wandered, aimlessly, helplessly.

There was nothing to do against the might of the Susquehanna River, which yesterday afternoon broke into the Knox Coal Co. Mine at Port Griffith.

Forty-five men were working in the tunnels when the swirling water ate a hole in the bank of the river and flooded the mineshaft below. Seven got out right away through an air shaft. But the flooded tunnels hindered rescue attempts and the people gathered to wait.

### Peer Into Ambulances

They peered anxiously into ambulances that brought the first seven survivors back to the company office, but the men were rushed inside to be questioned about the rest of the crew.

At Pittston Hospital many waited. "Any word of my brother?" asked one pale and bewildered woman. "I'm trying to find out whether my brother got out of the mine.

Do you know whether there are any men still in there? Do you have the names of those who got out?"

One could only shake his head. At the time 38 men were still in the mine.

Then early last evening the hospital's switchboard operator received an exciting call. "They have found 35," she shouted.

Nurses and doctors rushed into the corridors, carrying medicine, blankets, pushing stretchers.

### Told to Receive Men

"The police just called and told us to get ready to receive 35 men from the mine," one nurse told a reporter.

About 7 p.m. an eight-man rescue crew had come upon 26 miners—not the reported 35 wandering in waist deep water through the mine, not far from the air shaft.

"We had walked and walked and walked for seven hours," one of the men, John Gustitis, 25, of 911 Tunkhannock Avenue, West

Fig. 31. Onlookers stand watch at the Eagle Air Shaft, January 22, 1959. (Courtesy Stephen Lukasik)

Pittston, said later. "We were sure happy to see those guys."

One by one the men, and finally the rescuers, were helped up the steep 35-foot [50-foot] air shaft with the aid of ropes lowered from above.

At the hospital, the corridor leading from the dispensary quickly filled with people. As each group of men was unloaded or stepped out of the ambulance, relatives and loved ones looked for a familiar face.

One grimy miner walked through the door. Two or three men grabbed him and hugged him.

### Tears Ran Down Her Face

"Hey, Al, congratulations, boy," someone shouted. A woman ran up to him and threw her arms around his shoulders. Tears ran down her smiling face.

Al was hustled into a nearby room where nurses and doctors examined him.

Inside the room, a nurse handed Gustitis a cup filled with orange juice. "What is this, beer?" he asked. Without waiting for a reply, he gulped a drink.

"Thank God, thank God," cried another woman as she recognized a familiar face among the survivors.

Luckily Joseph Stella, 35, of 325 Market Street, Pittston Township, a mine surveyor, was with some of the seven men who escaped right after water started coming in. Stella is chief of police of Pittston Township.

### Map of Workings

"I had a map of the mine in my hand when the water came through," he said. "Me and some others followed the map to reach the air shaft."

Another of the original seven was Joe Soltis, 43.

"It just suddenly came in on us," he said of the break. "At first I was panicky, then I thought about the air shaft, and headed for it and got out."

But off to one side of the hospital corridor, a young woman buried her face in a handkerchief as the last of the 26 men came in. Somewhere underneath Port Griffith 12 men who had gone to work in the mine yesterday morning were either trapped or drowned.

### News Story: Two Rescuers

The stories of rescuers Anthony Arnone and William Hague (figure 32) are presented in this news account. (From "Rescuer Tells Story of 24 'Happiest Guys,'" Wilkes-Barre *Sunday Independent,* January 25, 1959)

Two of the first rescuers who appeared at the Pittston Hospital Thursday night, after the large group of survivors were sighted, had a stirring tale.

They are Anthony Arnone, 38 East Oak St., Browntown section of Pittston, and William Hague, 2 Wilkern St., Pittston, who told of the dramatic moment when the group of 24 [26] men were located.

Coated with muck and grime, the clean blankets around their shoulders incongruous, the pair sat in the emergency ward of the hospital sipping black coffee and dragging deeply on cigarettes.

This was at 7:30 P.M.

"We went down there (the air shaft) about 5 [p.m.], I guess," Arnone said.

"I was on post about 400 feet from the opening of the Eagle air shaft. Another man was about 50 feet away on my right. There was a concrete wall with a small opening between us," he recalled.

### HEARD A SHOUT

"Some of these boys (the survivors) must have seen the other fella's light. Anyway, I heard a shout and at first I thought it was the other men on post. We heard the shout again and we hollered back.

"Pretty soon we got sight of the guys. . . . I think there were about 20 or 24 in the bunch. I was on one of the highest spots. There wasn't any water around there then," Arnone continued.

"But then we saw them. Those guys were the happiest I've ever seen.

Fig. 32. William Hague, miner, Knox Coal Company.

"Some of them were still praying. A lot of them were crying. One of my buddies grabbed me and said, 'You little son-of-a-gun; I've never been happier to see anyone in my life.'

"That air shaft (the Eagle) is about 75 feet deep. It's a miracle any of those guys got out," Arnone said. Arnone recalled the names of some of the men in his rescue team. Wearily, he listed these names, [William] Hague, James Jamieson, James Jeffries, William Receski and Joseph Hopkins. He said he couldn't remember the others right then.

His hands shook as he lifted the cup of strong black coffee to his lips and the crisply starched nurses in the emergency room bustled around him.

### Oral History: Violet Williams

Violet Williams' (figure 33) position as the night supervisor of nursing at the Pittston Hospital placed her in the center of activity when the survivors arrived for care. (From Violet Williams, taped interview, May 9, 2003, NPOLHP)

VIOLET WILLIAMS: Well, it was bedlam at first. They didn't all come at once, a few at a time, so we were able to take care of them in the emergency room and then they were transferred up to the ward. There were no serious injuries. They were mostly [suffering from] exposure and some of them had breathing problems. The doctors in the emergency room examined [them]. The workers' vital signs were taken and then they were taken up to the ward where they were cleaned up and given warm clothing because they were wet.
ROBERT WOLENSKY: Was there any hypothermia?
WILLIAMS: Not really. Nothing severe. They were given warm drinks and something to eat and they were very happy to be out of the mine. Their people were there and the clergy were there. They really gave a lot of comfort. . . . The families were able to go up to the ward and stay with them and they could visit them until it was bed time. . . .
WOLENSKY: Did you bring in the off-shift nurses?
WILLIAMS: Yes, some were called to help. The day shift nurses stayed on. They stayed on longer than their [quitting] time. Then every-

Fig. 33. Violet Williams (foreground), former nurse, Pittston Hospital. (Courtesy of Willaim Best)

thing was very well under control. It was very orderly. We had plenty of help. As the men were treated and examined they were transferred up to the ward and the nurses in the ward took care of them. By the time they were [taken] up [to the ward] more would come in so they were spaced out enough that they were all well taken care of.

WOLENSKY: How were they brought to the hospital?

WILLIAMS: By ambulance. I imagine it would be Jenkins Township ambulances. They were all mobile, ambulatory, so they didn't need an ambulance for each person.

WOLENSKY: Any broken bones?

WILLIAMS: No.

WOLENSKY: Any cuts? Serious cuts?

WILLIAMS: No, it was mostly exposure.

WOLENSKY: What are the symptoms of exposure?

WILLIAMS: Cold, chills.

WOLENSKY: Fatigue?

WILLIAMS: They had fatigue I think from being in the mines but they responded quickly. When I made rounds again in the ward everybody seemed to be happy and glad to be warm.

WOLENSKY: Pretty good spirits up there?

WILLIAMS: Yes. . . .

WOLENSKY: One survivor said that somebody gave him a shot of whiskey!

WILLIAMS: Could be, because miners liked whisky. They used to say that it cut the cold down. Maybe a relative [gave him the whiskey]. He was cold probably and someone [gave it to him].

## Oral History: William Hastie—The Tragedy of Benjamin Boyar and Francis Burns

Knox laborer William Hastie uses his extensive mining expertise to examine the demise of two victims who were repairing pumps in the lowest reaches of the operation, the Red Ash Vein. (From William Hastie, taped interview, July 31, 1989, NPOLHP)

Frank Burns and Ben Boyar were completely out of touch. There was no way of warning them and they were doomed from the beginning. False bottoms had been placed in the shafts at the Marcy Vein [second] level so the hoisting cages no longer served the bottom veins. We don't know when these men encountered the water. We know that it probably filtered through to the lower veins more slowly. Access to the bottom veins was by what were called man-ways. They were small openings down through the rock just large enough to accommodate a ladder or perhaps a tight spiral staircase. It wouldn't take much water to plug them up.

We don't know when they encountered the water. Maybe it found them where they were working or maybe they met it on the way out when they began to climb up through a man-way. We can imagine their puzzlement at first and then their horror as they realized what was happening. They would probably scurry around trying to find a way out. The water would accumulate more slowly in the bottom vein and they probably took a long time to die. Their death was probably slower and more horrible than any of the others and we can only hope that they had heart attacks or something like that from the cold water and from the fear.

## News Story: Missing and Rescued

Following is a list of the missing and the rescued containing most of their ages and addresses. ("The Missing and the Rescued in Mine Flood," Wilkes-Barre *Sunday Independent,* January 25, 1959)

### The Missing
John Baloga, 54, 72 Race St., Port Griffith; William Sinclair, 48, 191 Vine St., Pittston; Daniel Stefanides, 33, 68 Lackawanna Ave., Swoyersville; Samuel Altieri, 62, 63 Reynolds St., Hughestown; Francis Burns, 62, 19 Elizabeth St., Pittston; Eugene Ostrowski, 34, 7 College St., Nanticoke; Joseph Gizenski, 37, Hunlock Creek; Charles Featherman, 37, Koonsville Rd., Muhlenburg; Herman Zelonis, 58, Pittston; Frank Orlowski, 42, 147 Main St., Dupont; Benjamin Boyar, 55, 30 Wesley St., Forty Fort; Dominick Kaveliski, 52, 8 Wood St., Pittston.

### The Rescued [from the Eagle Air Shaft]
George Mazur, 23, 1114 Exeter Ave., Exeter; John Pientka, 51, 377 North Main St., Plains; Louis Marsico, 220 Pine St., Old Forge; Myron Thomas, 43, 402 Union St., Taylor; Anthony Krywicki, 53, 1368 Main St., Port Griffith; John Gadomski, 31, 47 Dorrance St., Wyoming; Stephen Cigarski, 42, 1242 Main St., Port Griffith; Joseph Francik, 39, 230 Mill St., Pittston; Merle Ramage, 39, 214 Elizabeth St., Pittston; Joseph Solarczyk, 51, 228 Orchard St., Exeter; Joseph Stella, 35, 325 Market St., Pittston Township; Frank Ludzia, 38, 104 Front St., Pittston; John Gustitus, 39, 711 Tunkhannock Ave., West Pittston; Paul Cawley, 42, 1623 River Rd., Port Blanchard; Joseph Kachinski, 42, 24 Griffith Lane, Wilkes-Barre; George Shane, 49, 50 Wood St., Inkerman; Edward Borosky, 39, 222 East Grove St., Taylor; Charles Proshunis, 54, 252 Battle Ave., Exeter; Joseph Moore, 42, 353 Tedrick St., Sebastopol; Angelo Retondaro, 51, 10 Defoe St., Pittston; John Blara, 47, 200 Susquehanna Ave., Wyoming; Francis Wascalis, no age, 111 Grove St., Exeter; Stanley Roman, 31, 1930 Susquehanna Ave., Exeter; Martin Saporito, 49, 107 John St., Pittston; Albert Smelster, 50, 31 Welsh St., Pittston; Michael Mackachinus, 43, 10 Union St., Inkerman; Charles Michulis, 51, 424 Gravity St., Pittston; Louis

Fig. 34. Crowd greets James LaFratte after his rescue from the Eagle Air Shaft. (Courtesy of Stephen Lukasik)

Randazza, 354 Parsonage St., Hughestown; Jerry Stuccio, Pittston; James LaFratte, 227 McLane St., Dupont [figure 34]; John Elko, 51 Butler St., Wyoming; Joseph Soltis, rear 1352 Main St., Port Griffith; Amedeo (Paul) Pancotti, 180 Schooley Ave., Exeter.

### Editorial: Afflictions and Questions

Within hours, editors began writing about the consequences and causes of the catastrophe. ("The Missing Miners," *Scranton Tribune,* January 26, 1959)

> The fate of 12 men missing since waters of the Susquehanna River burst into a mining operation at Port Griffith Thursday remains the paramount consideration among all of the ramifications of the tragic event. Hope for their eventual rescue, while it diminishes with each

passing hour, will persist, no matter how faintly, until the last reason for hope is gone.

Early reports yesterday that workmen virtually have stopped the flow of water into the mine were gratifying. Reduction of the flood condition will permit rescue crews to go underground for a search for the workers who were cut off by the rush of water.

Diversion of the river away from the mine also should halt the underground damage and the flooding of other mine operations. While there is reason to believe that some of the pits which must have been flooded may never be able to be worked again, the elimination of the source of the flooding could prevent the spread of irreparable damage and ease the threats of a drastic and crippling effect on the anthracite industry.

The great need for federal and state assistance in this situation is readily apparent. Prompt pledges of federal and state aid were welcomed and it is incumbent that there be no delay in undertaking assistance measures. The possibility that some $10 million left from a $17 million fund appropriated for mine drainage work in 1955 might be available for use in the present crisis should be explored immediately.

Whatever legislative action is necessary, either in Washington or Harrisburg, should be started.

Meanwhile, all in this region share the anguish and grief of the wives, children, other relatives and friends of the missing miners. Their trial is great and their perseverance, faith and courage are touching. Any step taken to help them deserves full support.

And certainly not to be overlooked, or unduly delayed, is the importance of an intensive investigation of the circumstances surrounding the tragedy. As the *Tribune* said Friday, "The mining practices and procedures at this operation should be minutely explored particularly in relation to the hazard allowed to exist by reason of the proximity of the operation to the river. Obviously, there was insufficient barrier to prevent a break-though of the river, an ever-present danger." (figure 35)

## Notes for Chapter Three

1. Contrary to the generally accepted view—including the account in Ellis Roberts' *The Breaker Whistle Blows* (Scranton: Anthracite Press, 1984, p. 149) and our account in *The Knox Mine Disaster* (1999, p. 20)—Mr. Thomas here suggests that the group directed by Pennsylvania Coal Company surveyor, Joe Stella, led the way to the Eagle Air Shaft, rather than followed Thomas' group. Similarly, in another news story in the *Scranton Times* on January 23, 1959, probably drawing upon the same interview used here, Thomas is quoted as saying, "It was then that I noticed that there were seven men missing. That last few men in the line of march told me that some men stopped to rest and could not keep up with the fast pace [of Stella's lead party]. Going back down the gangway a bit, looking and shouting their names, I could not find them. Returning to the main body of men, I asked if Joe Stella was with the missing men. When informed he was with them, I was more at ease, knowing that he would lead his men out." While the oral history quoted in this chapter by John Gadomski, and oral histories with Myron Thomas' son, Robert (May 28, 1996) brought the controversy to our attention during research for our 1999 book, we lacked sufficient corroborating evidence. However, we have since uncovered additional oral history and newspaper documentation (including this news story and a radio interview with Myron Thomas) that supports the credibility of Thomas' assertions. Still, the rescued miners provide different interpretations of the separation, as apparent in the oral histories in this chapter by Edward Boroski, Joseph Francik, George Mazur, and Stanley Roman.

2. Because of the oversupply of certified miners in anthracite, many had to accept jobs as laborers in order to earn a living. Gadomski held the top position of miner on his crew while Roman and Gustitas, even though certified miners, worked as laborers for lower pay.

Fig. 35. Two rescuers wait at the Eagle Air Shaft. (Courtesy of Stephen Lukasik)

# Chapter Four
## Spouses and Siblings

*I awakened my husband and related my dream. I kept saying, "One of my brothers is going to die." He tried to assure me that it was only a nightmare. It was a nightmare that turned to reality that very same day.*
Frances O. Sinclair, sister of victim Frank Orlowski

*It's been so long each day. I go out to the porch and watch what is going on at the place where the river broke in and see everything that goes on. So you see it's very hard for me to bear. He is so close and yet I can't go to see or talk to him, but I know he is watching over us at home, from God's Heaven.*
Caroline Baloga, wife of victim John Baloga

The twelve victims of the Knox mine disaster left behind forty-two spouses and children and numerous brothers, sisters, parents, and other relatives and friends. The stories of the victims' families represent a second important set of voices related to the human ramifications of the disaster. This chapter presents the views of spouses and siblings (except the last interview with Ida Gizenski which includes her two sons), while

Fig. 36. Onlookers gather near the River Slope Mine in Port Griffith. (Courtesy of George Harvan)

the next chapter presents the children's recollections. The memoirs are augmented by accounts from local newspapers.

## NEWS STORY: RELATIVES HUDDLE

Families gathered near the Eagle Air Shaft waiting for their husbands, sons, fathers, grandfathers, and brothers (figure 36). (From "Relatives Huddle in Small Shanty to Await News of Their Loved Ones," *Wilkes-Barre Record,* January 23, 1959)

> An icy wind whipped across the hill at Port Griffith which loomed over rescue workers as they worked frantically to save the trapped men and to stop the Susquehanna River whirlpool from gushing into the River Slope, Knox Coal Company yesterday.
> 
> As long as daylight prevailed, the curious swarmed over the hillside, ignoring warnings of the State Police to move back from the edge of the hill. Gravel knocked loose by the curious tumbled down on rescue workers.

Trucks dumped loads of excelsior atop the hill to be pushed into the hole, the bales silhouetted against the cold winter sky. Below could be seen the workmen, pulling exhausted mineworkers one by one from an air shaft.

Some relatives gathered at the air shaft, eager to be on hand when their loved one crawled to safety. Most remained in the shanty near River Slope, approximately 400 feet away, crowding in a small room heated by a tiny coal stove and finding seats where they could among the machinery in the shanty. Here a woman stood with her young child in her arms and there two or three men, dressed in working clothes, spoke quietly about the disaster. In a corner, seated on the crude wooden bench, a group of women sat silent, some with wet eyes and others staring into space. In this building, the full impact of the tragedy could be felt.

The cry of "They're bringing one out," signaled a rush from the building and each rescued miner was surrounded by anxious relatives seeking word about the others.

"Who is it?" one woman anxiously queried from the edge of the crowd, pushing her way to the rescued miner. Another grabbed him by the arm and asked if he had seen her husband. "I don't know anything. Let me through," the miner replied, the horror of his experience written on his face.

One elderly miner cried like a baby when he was pulled to safety from the air shaft. Most said nothing. Trundled to the shanty by jeep over a railroad bed, the only way to get to the scene of the rescue shaft, the men were placed in ambulances at the shanty and then taken to Pittston Hospital, a few blocks away.

Corridors at the hospital began to fill with relatives and friends when word spread the miners had been rescued. Extra nurses, doctors, and office staff worked frantically to give first aid treatment to the men and admit them as patients. Huddled in blankets, the exhausted, grimy miners were pushed in wheelchairs one by one to the hospital wards. Relatives walked with some, not wanting to leave the side of their loved one. Smiles of greeting were exchanged here and there by friends in the corridors.

There was no happiness for some as they tearfully asked the information desk if their father or husband were in the hospital, only to be told he was missing. One student nurse, Theresa Orlowski, waited anxiously as each ambulance came in, looking for her father, Frank Orlowski. She collapsed and was admitted to the hospital infirmary when she learned he was not among those rescued.

As the men settled down for their needed rest, the corridors clear, relatives returning home happy that their loved ones had been rescued but saddened in the realization that other men were still missing.

As one haggard wife cried, "Thank God it was no worse. It might have been such a horrible disaster."

### News Story: Hopeful Families

Three reporters visited the Kaveliskie family and tell their story. (From Libby Brennan, Tom Moran, and Tom Heffernan Jr., "Rays of Hope Found in Homes of 12 Men in Flooded Mines," *Wilkes-Barre Sunday Independent,* January 25, 1959) (figure 37).

Fig. 37. Pastor greets rescued mineworker at Eagle Air Shaft. (Courtesy of Stephen Lukasik)

# He Didn't Have a Hunch
## The Koveleski [Kaveliski] story is a different one.

Dominick Koveleski, 52, of 8 Wood St., Pittston, one of the trapped mine workers in the Knox workings, had many premonitions or hunches in his years around the mines. Every time he felt a "hunch," he would remain from work.

Last Thursday Mr. Koveleski failed to have that "feeling or hunch" as he would call it. He left for work and last night he was still trapped in the workings holding 12 men entombed in the dark caverns of the mines.

Mrs. Koveleski, with her son, Dominick Jr., his wife and a neighbor, greeted a reporter anxiously. They feared bad news but were quickly assured the news they feared most was not forthcoming.

### Married 28 Years

"We have been married 28 years," Mrs. Koveleski said, and added, "Dominick was 52 on New Year's Day. He has worked around the mines practically all his working life and was at the Pennsylvania [Coal Co.] mines for 10 years."

"One of the rescued men visited us today," Mrs. Koveleski continued, "and told us he could not understand what happened to my husband. I only know the miner as 'Big Zack.' He told us when the trouble started, he told my husband that he would lead the way because he was much bigger than my husband. He started to lead and when he looked around again, my husband had disappeared. Timbers, ice and high waters were then rushing in."

### Many False Reports

Mrs. Koveleski like so many of the families has been getting many discouraging reports, all of them false. The shock has been great and Mrs. Koveleski required medical attention. "It's a pretty rough situation," Mrs. Koveleski remarked. But she was fortified by her son and daughter-in-law who are steadily at her side. Dominick Jr. is employed at New Jersey but raced home when the news broke.

Mrs. Margaret Chescavage, landlord of the Koveleskis, was also grieved at the misfortune. "Mr. Koveleski is a good man," she went on, "always ready and willing to help me with my property. It's really too bad."

Despite all the harrowing reports, Mrs. Koveleski and her little family have strong hopes and faith in their God that Dominick will be returned safely.

Nearby on the street, men asked, "Why don't you get the real story about working in the dog hole? Do you know what a dog hole is? Why were the men warned at 1 o'clock, two hours after the first warning came in?" These questions filled the air as the reporter went on to another home bearing the sorrow of the current tragedy.

### NEWS STORY: VICTIMS' FAMILIES TROUBLED

Anxious and angry family members waited with hope. (From "Families Irked by Lack of Official Word from Coal Company: Tired, Tearful, Tragic Wives of Trapped Men Praying, Hoping, Anticipating but Fearing," *Scranton Times,* January 25, 1959)

The wife of a miner lives in terror—and in hope.

That was apparent yesterday as this reporter visited wives of three of the 12 miners trapped in the flooded River Slope of the Knox Coal Co. operation in Port Griffith.

The women—tired, tearful and tragic—have been waiting, imagining, fearing, and anticipating for three long days.

When they bade their husbands goodbye in the pre-dawn darkness Thursday, they had no idea that midway through the shift they would hear of the disaster.

And all three expressed astonishment and anger at not being notified by the company—even until late Saturday afternoon—that their kin is among the missing.

"If we hadn't heard it on the radio and read in the paper, we'd never know why our husbands didn't come home Thursday night," was the stock complaint.

While hope for the safety of the miners may be waning in the public's mind, the families of the loved ones have indestructible faith.

Mrs. Frank Orlowski, 147 Main Street, Dupont, was resting on the divan when we called. Swollen-eyed and dazed, she apologized for her appearance and the chill in the house.

"Since Frank isn't here to take care of the furnace, Bobby (their

13 year old son) has been building the fire, and he just can't seem to get the coal to burn," she explained.

"You're the first outsiders to come," Mrs. Orlowski revealed. "The neighbors and relatives have been wonderful, but we haven't heard a word from the company."

Recalling the events of their last morning together, she said: "Frank never eats much breakfast, just coffee. But that morning we were up at 5 a.m., and my husband said he was hungry, so I fixed him some cereal. Then he had two more cups of coffee.

"It all seems strange, now that I look back. He even asked me to fix a couple of extra sandwiches for the pail. I only hope he's not starving to death now," she said as her voice choked with emotion.

Frank, who is 41 [42], has spent most of his working years in the mines, with the exception of the four years he spent in service. Mrs. Orlowski is accustomed to the lives led by miners' wives. Her father spent 40 years in the hazardous occupation, and she reminisced about "how my mother always worried about him."

"That's his (her husband's) car down there," she said, pointing out the window to a light blue freshly polished automobile. "We haven't had it too long. When my niece saw it at the mine that afternoon (of the tragedy) we were sure Frank must be safe, because the door was open.

"We thought he must be working on the rescue crew and must have gotten out of it in a rush. But he must have been trapped, because I talked with his miner's wife (Frank was a laborer) and she told me her husband Frank was right behind him when the gang started out," explained Mrs. Orlowski, the former Irene Exeter, a native of Dupont.

The interview was interrupted at this point by the arrival of the couple's daughter, Theresa, 19, a student nurse at Pittston Hospital where she has been confined to the infirmary since the disaster.

The meeting was the first between mother and daughter since the flooding of the mine, and the scene was a heart tugging one.

"Theresa and her dad were very close," the mother disclosed. "He planned to teach her to drive this Spring," as she broke down some more.

Between sobs she told how "Frank always went to the hospital to pick her up on her day off. He was so proud of her."

"We would have been married 20 years in April," concluded Mrs. Orlowski.

Mrs. Samuel Altieri, 62 Reynolds St, Hughestown, answered the door herself. The living room was filled with family and friends—sitting around waiting for word. They had already made two trips to the mine but were unable to obtain any information.

"They said now that the Army had taken over, they weren't allowed to give out any news," said one of the daughters, Mrs. Herman Ciampi. "Can't you put something in the paper about the company not even notifying the families about their men?" she asked. "Tell them how you found out," she said to her mother.

Mrs. Altieri, whose husband 65 [62], is an electrician, disclosed the following story:

"Sam and I always came home on the same bus at 3:15 p.m. (She works at the Pittston Apparel Co.) I'm already on it when he gets on. But Thursday he wasn't on the corner where he always waits.

"I didn't get excited because once in a while he gets a ride home. Then we heard sirens and when the bus slowed down we saw ambulances going by.

"The lady who sat with me said, 'Wasn't that a terrible accident at the mines?' I didn't know anything about it. I was at work that day. 'What mines?' I asked, and she told me.

"But I didn't think Sam would be down that day. Electricians don't have to go into the mines every day. But when he didn't come home, we went down to the slope.

"When we couldn't find him there, we went to the hospital. Oh, it was terrible there—everyone searching for the familiar face of a husband or father.

"But the men who know Sam told us that he must be safe. 'If anyone knows the mines, he does,' they said, and several of them had seen him. He told us not to get excited but to hurry and get out because there was going to be a flood,' the rescued miners told us," explained Mrs. Altieri.

Mrs. Frank Ferrara [Ferarre], another daughter, volunteered: "The only thing we're afraid of is that water rushed him from behind."

"If I know Dad," added young Sam Jr., 14, "he's up high somewhere, watching the water go by."

Mr. Altieri, too, comes from a family of miners. His brother, James, was killed at the No. 6 Colliery in 1950. Their father carried a severe scar on the head—incurred in a mine accident—to his grave.

Another son, Vincent, who teaches at Oklahoma State University, was en route home by plane after the family notified him Friday. The trapped miner's other daughter, Mrs. Philip Aiello, Cleveland, Ohio, learned of the disaster Thursday night while watching a television news program. She phoned immediately and when informed that her father was among those missing, flew home yesterday.

Mrs. Alex Parente, the last of the six children, resides in Cuba where her husband is serving with the Navy. She is unaware of her dad's plight, although the Red Cross has been asked by the family to contact her.

"How long do you suppose their lamps will stay lit?" asked Mrs. Altieri, her anxiety giving way to tears. "They must be in total darkness by now. How could they ever find their way out?" she asked plaintively.

Sam had talked often of retiring and spending his time in the garden, "if only he had!" she concluded.

While the visits to every home were heart-breaking, a pair of gray work gloves on top of the refrigerator at the Baloga home—just 100 feet from the ill-fated Knox—was the saddest.

"There are his new gloves up there," beckoned Mrs. John Baloga, 12 Rose St., Port Griffith. "And he'll never get to wear them."

Her 54-year-old husband and father of her four children had asked that she buy him a new pair. "You can have them to play with, Johnny," she said to her six-year-old, "Daddy won't need them now."

Mr. Baloga, a veteran of 30 [35] years' employment in the anthracite mining fields, is in the second stages of miner's asthma.

"The doctor told him he shouldn't work anymore, but he said he wasn't too bad, and just kept on struggling," said his wife, bursting into tears.

While her daughters tried to console her—their own eyes red with weeping—she twisted a handkerchief in her nervous hands and related more details of her life with her husband.

"He was such a good man," she said. "He worked and worked

and never spent a nickel foolishly. He just brought the money home to me and never asked how I spent it.

"And he was so good to the children—he'd never hit them. I don't know how I'll live without him now."

At that point representatives of the Salvation Army called to offer consolation, as well as any material help which might be required. Then, before leaving, they offered prayers for the family.

### Personal Letter: Frances O. Sinclair

Looking back after forty-one years, Mrs. Sinclair remembers her brother, victim Frank Orlowski, and describes two apparently accurate premonitions regarding his ruin. (From Frances O. Sinclair, sister of victim Frank Orlowski, to Robert P. Wolensky, January 17, 2000, after reading the Wolenskys' book on the *Knox Mine Disaster* [1999]. Mrs. Sinclair was not related to Knox victim, William Sinclair)

Dear Mr. Wolensky,

I wish to thank the Wolensky family for the dedication and the work in accurately depicting the disaster at the Knox mine on January 22, 1959, where my brother Frank Orlowski perished. It took me a while to have the courage to read the book and to relive the pain of that dreadful day.

My dear friend, Dolores Tomaszewski Jones, attended the [fortieth anniversary] Mass and Dedication at Port Griffith [on January 24, 1999] and obtained the book for me signed by you and your brother Ken and daughter Nicole.

In November of 1958 my mother and my family were invited to spend Thanksgiving holidays with my brother Frank and his family in Dupont. We had never been together for Thanksgiving before. In the evening after all the children were in bed, my [son] Tom, Frank, my husband Don and I sat around the table and talked late into the night. It was then that Frank talked about his work and the conditions at the mine. In talking about the dangers at one point he said, "One of these days we are all going to die with our boots on like rats in a trap." It was exactly two months before the disaster. Those words came back to me on January 22, 1959, when my husband came home for lunch and told me of the phone call that he received at the office: "the Susquehanna River flooded the mine where Frank works." I almost collapsed, my knees buckled, and all I could say was, "Oh, my God."

My Mom was in California visiting my sister Helen and her family. It was early morning in California. Mom and Helen were having breakfast and Mom was telling Helen about a dream she had that night that disturbed her. She dreamt that a gardener was cutting down a healthy, beautiful tree laden with fruit. In her dream she said, "Why would anyone cut down a tree with fruit on it?" She then said to Helen, "Someone is going to die." At noon Helen's husband came home and told Helen of the phone call he received at work. My brother John made that call. He didn't want to break such devastating news to Mom and Helen without the support of Helen's husband being there. My Mom flew back to Dupont that very same day.

When Frank was 17 he enlisted in the Army and spent three or four years in the Hawaiian Islands stationed at Schofield Barracks. He loved Army life and often told us stories of his experience, the land, and the people. Sometimes he would put on the grass skirt he brought back from the Islands and entertain us with his version of the "Hula Dance." It was always very funny. He liked to make us laugh.

On January 21, 1938, while in Hawaii, Frank was notified of his sister's death. Agnes was 25 years old when she died of pneumonia and left a husband and three small children. Frank was granted leave but he did not arrive in time for her funeral. It was exactly 20 years and one day later that Frank died in the mine.

I would like to relate a dream I had at about 3:00 a.m. on the day of the mine disaster. In the dream my husband and I were attending a party and dance at his dental fraternity. We were on the balcony watching the dancers below. A man, while leaning over the railing of the balcony, fell to the floor below. I grabbed my husband's hand and said, "Let's go and see what happened." When we reached the floor below it was no longer a dance floor but a mine with downed pillars and huge black boulders. The man was lying face up over a fallen pillar, his chest was ripped open and blood was spurting from it. I awoke in a panic. I awakened my husband and related my dream. I kept saying, "One of my brothers is going to die." He tried to assure me that it was only a nightmare. It was a nightmare that turned to reality that very same day.

Frank's daughter, Theresa, was a student nurse at the Pittston Hospital. She watched the rescued men being brought in and collapsed when she realized that her father was not among them. His son, Francis Robert, "Bobby", was 14 years old. . . . I am 80 plus and the last of the Orlowski siblings.

I appreciate the inscription on the fly-leaf of your book, "to Frances Orlowski and the Sinclair family, Best Wishes." Again I thank you for the great work on the book.

Sincerely yours,
Frances O. Sinclair

### Oral History: Lea Stark

Lea Stark (figure 38) lost her husband, William Sinclair, in the disaster. A native of Scotland, he was the nephew of another Scottish native, Knox superintendent Robert Groves. Mrs. Stark recalls the loss of her husband and the resulting anger over Groves's decision not to notify the workers that the river had broken in. (From Lea Stark, taped interview, September 14, 1990, NPOLHP)

LEA STARK: I didn't want him to go to work [that day]. Now wasn't that funny? There was an awful bad storm [the night before], a windstorm. I'm afraid of wind. I think ever since then I've never been afraid of wind. It was an awful bad storm and I said to him, "Why don't you stay home [tomorrow]?" Then I said, "Well, see how it is in the morning." And here it [the disaster] happened in the morning. You know, it was calmer and the weather wasn't so bad. I had no feeling [that something might happen], but it was just that because of that windstorm I was afraid. . . .

I had been ironing and my girl friend called me up and said, "Lea, lets go down to the movies." I told her I was ironing. She said, "Oh, come on." So I left a note for him that I had gone with my friend to the movies. I didn't know anything about it until we came up through Port Griffith. I could see everything [police, ambulances,

Fig. 38. Lea Stark, widow of victim William Sinclair.

etc.] around and I knew something happened, but I didn't even bother [to stop]. When I came home his uncle was here and he said, "Where have you been, Lea?" I said that I went to the movies.

See, he only had one uncle here [because] he was born in Scotland. . . . Billy came over here when he was a boy. He was a little older than me. He came to [stay with] his aunt. He left his mother back there with that father he had. He came out to this aunt and uncle. . . . I was born in Wales. Yeah, we were always gonna go back together, you know. My grandmother kept me. My father died when I was a baby. My grandpa and grandma raised me and then my mother married again and [later] she came [to get me] but they didn't want me to go. They wanted me to stay with them. I came to this country when I was twelve years old. In November, when I think of it now. The weather, Oh God [bad]! I came all by myself. My grandfather showed me off to Southampton. My mother met me in New York. Twelve years old.

I didn't go down [to the River Slope]. I just didn't feel I could. Thank God my mother was alive then. She lived close to me and she came down and we sat up all night. Bob Groves, I guess you heard about him? Scotsman from Scotland. He was my husband's uncle and he went right down [into the mines]. He knew where Billy was, you know. He was the superintendent and he went down in the water [which] was coming just like that.

Mike [Lucas] got out. You know they didn't tell them the bottom was cut [out of the river]. They didn't tell them the river was coming in. That was the bad thing. I said to Bob, they were up here, and I said, "Look-it, Bob . . ." He said, "Well, we didn't want to scare them, Lea." I said, "Scare them?! My God, my husband would have been out." I said, "He would have been out, with no doubt!"

Oh no, they weren't [going to tell them]. They never told them. They should've told them. Him and Stefanides. He was with Stefanides who was only a young fellow; he was younger than my husband. . . . They could've been out. Mike got out. Mike was a miner. They hollered to Mike to come. Mike came down and went out. I know Mike feels bad about it. But see, he didn't know either. They weren't told, they weren't told. Bob told me himself, they didn't want to scare [them]. I said, "Scare them, my God, it was their life!"

Everybody was here [after the disaster]. It was hectic for a long time. I mean people were so good to us. They were wonderful. They'd come and they'd come. [Congressman] Flood wrote me a letter. There were fundraisers for the families. They put on dances and things. Some of the factories collected money. They were real good. The people were very, very good to us, very generous. They were wonderful, the people were. . . .

ROBERT WOLENSKY: Did Billy like his work in the mines?

STARK: Yeah, well with money, see, that was good money, big money. . . . They were making better money than he ever, ever made in the mines. . . . They didn't realize the danger that they were so close to the water. They never should have let them work there. He never said anything about it. He never complained about the Knox Company. He worked [for] a lot of other companies. In fact, it was Bob Groves that got him this job with the Knox. He was at Dial Rock in Wyoming and that closed down. Then he went to Courtdale and then Bob got him this job in the Knox.

He started [in the mines] when he was young. His uncle came to this country and he was in the mines. My husband started as what they call a door tender [nipper]. He was young. I don't know how old he really was when he came here. That's all he did here was the mines. He was a laborer. Mike was the miner, see. He also worked with a Stefanides brother, Joe. Joe Stefanides felt terrible. Oh, Joe felt terrible over the loss of his brother.

I never talked to the Knox officials. Mike came over. I'm not sure I wanted to talk to him, but I don't think any of the others [came over]. I don't know exactly what Mike said; you know, he felt so bad. Oh, he felt terrible. I know he did.

I was angry. I was angry at Bob Groves. I sort of resented that because they shouldn't have been working there, but then what could you do, you know? What it was that got me, they didn't tell them that the river was coming in. That's the whole thing. That was terrible. . . .

They put a memorial down in Port Griffith. By the church there they erected this monument. I gave a hundred dollars towards erecting it. All of us did, I guess. . . .

WOLENSKY: Did you sue the company?

STARK: Yes. I mean, I didn't ever get anything [much]. I don't know if everybody sued. We sued and we got $20,000. We didn't go to court. We got the lawyers, Alan Kluger and George Sporher, and they got it. They didn't have much trouble. It was quite a while [after my husband's death]. . . .

WOLENSKY: Did it take you a long time to recover from the loss of your husband?

STARK: Oh yeah, yeah, oh yeah. I couldn't stand the house. I couldn't stand my house. I was out all the time visiting somebody, going some place. I couldn't stand the house. It took me a couple of years. He was a good guy. Everybody liked him. He was a nice guy. Good in his church. He was head usher in the church. He liked to go to church. I don't know why it had to happen. They say how things happen, and to think what happened to him. Someone should've been telling him, "Get out, get out, get out!"

## ORAL HISTORY: OPAL FEATHERMAN

Opal Featherman's husband, Charles, perished at Knox. Vivid memories of his final morning at home have stayed with her (figures 39 and 40). She also describes the difficult aftermath. (From Opal Featherman, taped interview, June 10, 1997, NPOHP)

OPAL FEATHERMAN: I was at home and I heard it on TV but I didn't really know that that was the mine he was working at. I knew that it was up there near Pittston, but I didn't know if it was the right one. Mrs. Gizenski's oldest son, his grandfather, and another boy came down and told me that my husband was trapped in the mines. That's how I really knew that they were in there. But I knew something was wrong because he didn't come home from work that night. They were working day shift and he always came home. I mean he wasn't one to stop off, you know what I mean, go to a beer garden or anything like that, so I knew something was wrong. I'll tell you that was an awful day. You couldn't believe. Then the next day, there was this guy [who] came around, if I remember right, from the mines and he had a priest with him.

My husband never was one to talk much but at Christmas time he said there was water seeping in right behind where they were working. In January it was real cold and he had this heavy coat that he wore. He took it off to work and hung it up. When he went to put it back on it was all covered with icicles, where the water had seeped in and froze. Icicles all over his coat! So I think they summed it up and surmised that something was gonna happen there. I don't think he was scared because he wasn't the type. He wasn't anyone to say any-

Fig. 39. Opal Featherman, widow of victim Charles Featherman.

thing like that, even if he was scared he wouldn't say so.

He worked with Mr. Gizenski. They used to take turns driving. And I knew Ida [Gizenski] for, oh, long time before that. We used to live right next door to each other when we were growing up. In fact, her sister married my brother. . . .

For about a week before, I had a funny feeling that something awful was going to happen. I really did. I told this before so it's not just when I'm talking to you, but I told this to people. I had this awful funny feeling and I knew something was going to happen, but I didn't know what. I really did. But I no idea it was anything like that. I just had that feeling in my stomach. I just knew something was going to happen. You know how you can just feel it? I knew something was going to happen, and it did.

And another thing. That morning when he got up to go to work, he got ready to go to work and he kissed me good-bye, and he went in and kissed Sherry [daughter] good-bye. She was still in bed because it was too early to get her up for school. He went in, he kissed her good-bye, and he came back and he just stood there looking at me, and he went back in and he stood there, looked at her, and he went to work.

Fig. 40. Charles Featherman in military uniform, circa 1944. (Courtesy of Opal Featherman)

And then he never came home.

ROBERT WOLENSKY: So it wasn't normal for him to look at you both like that?

FEATHERMAN: No, no, no. And that made me feel more like something was going to happen that day, and it did. . . . .

My daughter Sherry was only eleven when it happened. She flunked that year in school. The teacher flunked her. She was doing good in school and everything. I mean, not real good but she would have passed. Yet the teacher flunked her. The teacher wrote me a letter. It said that it was her nerves. That same night I took her right down here to the doctor. The doctor said, "I can't see anything wrong with her nerves." He said, "I'd go talk to that teacher." I went out to the teacher's house and I told her what the doctor said. The teacher said, "Well it's not exactly her nerves but she just don't seem to have her mind on her work." I said, "Well, my gosh, she just lost her father." I said, "What do you expect? You don't lose a father everyday, you know." But that teacher was heartless.

WOLENSKY: Did you sue the company?

FEATHERMAN: Well, it was never like a court case or anything like that. But I had a lawyer and they settled.

WOLENSKY: How long did it take to settle?

FEATHERMAN: (Laughs) Seven years.

WOLENSKY: Did you get any financial assistance in the meantime?

FEATHERMAN: Well, my daughter and I got Social Security. We survived. . . . But we missed him all the time, you know. You never get over it really. It was hard. He was handy at a lot of things. He was strong, big man. He was six foot tall and real strong.

## ORAL HISTORY: STEPHANIE STEFANIDES

Stephanie Stefanides, wife of Knox victim Daniel Stefanides Sr., remembers the early hours and days of the tragedy, as well as the long months afterwards. She raised four young children and made a new life for herself (figure 41). (From Stephanie Stefanides, taped interview, September 14, 1990, NPOLHP)

STEPHANIE STEFANIDES: My husband had a gas station beginning in forty-eight or forty-nine [and into the] fifties. Things weren't good so he gave it up. His brother [Joseph] worked in the mines and he was the one that encouraged him [to work there]. The job was there and he took it, although I didn't like it, you know. I didn't. His brother was hurt maybe a week before this happened. In fact, I think he was in the hospital at the time that this happened. My husband never worked in the mines before that. He didn't work all that long. He was [there] two-and-a-half years. That was it.

We were married in forty-eight. I graduated from GAR High School in forty-five and in forty-eight [we were married]. My mother died when I was very young. Just like my children were without a father, I was without a mother. But I raised the kids the best I could. They're all pretty good.

My one brother worked in the mines, Frank. My father-in-law did and my father worked in the mines. It wasn't the place to go. I guess my husband just thought, well he would go for a short while and then maybe things would pick up with jobs or whatever.

ROBERT WOLENSKY: The gas station wasn't making a good living for you?

STEFANIDES: It really wasn't. Even after he died there were people that owed him money. Bills were burnt and thrown away. I try to go for a walk every day and I pass by the place. We were there a lot with the kids. It's on the [Wyoming] Avenue. There's a florist there now, corner of Owens and the Avenue.

I don't think that he would have stayed [at Knox]. I doubt it. We were just too young at that time [and] it's not a place to be. But it didn't work out that way. People say, "What is meant to be is just going to be," accident-wise or health-wise, so . . .

WOLENSKY: Was the money pretty good at Knox?

STEFANIDES: It was better than the gas station. I'll tell you when he took the job in the mines, with the four children, it was hard for us from the beginning. Then we bought this home. But when looking back now, no, not really. It wasn't worth it, not at all. Of course, nobody deserves a death like that. I'm not only saying him, but the others too. That's not the way to go, you know. And so young—early thirties.

It was in the afternoon when we heard that it happened. I heard

Fig. 41. The Stefanides family, photograph taken immediately after the disaster. From left to right: Patricia (age five), Daniel Jr. (age nine), Mrs. Stephanie Stefanides, Michael (age one), and Christine (age two). (Courtesy of Stephanie Stefanides)

it on the radio. Christine and Patsy, two of the kids, were sleeping. Patsy was in kindergarten and Danny was in first grade. I can't even remember exactly. It was like an eerie, eerie feeling. When I heard the news flash I called his other brother and I said, "I think that's where they're working." He said, "No, I don't think so." I said, "I think so." Because [my husband] did mention the mine. He said, "Oh, I don't think." But later on it was, "Yeah." Well, we went up there. They thought that maybe there was hope, but . . . there were so many people [up there] and then we just came back home.

There were a lot of people coming in and out of the house. [We were awake] the whole night. I know my neighbor came and she stayed that night. My sister-in-law was here. But after a while everything just tapered off and that was it, you know, life goes on. . . . The days went into weeks and months and that's just how time has gone by. I don't know, the days just went by, the years [went by]. And, you know, raising four kids, it's [hard]. This is it. Everybody says, "Is it that long already?"

WOLENSKY: Did the company officials talk to you?

STEFANIDES: Not really. I don't know who came. I can't even remember who came here with the minister. There was really nothing from them at all. Nothing that I could really say. Not to me, maybe to the others. There was [Social Security] compensation so I got so many weeks [of income] at first, and then the children got it until they were eighteen. I did talk to a lawyer about suing but he told me I would get nowhere. He said compensation would be the best thing to do, you know, go that way. The four kids are small, he said, and over twenty years that would be the amount of money you would get. We had no problems at all with Social Security. The compensation was like 128 weeks were mine and then after that it was for the children until they were eighteen. . . . They did have a lot of fundraisers. They did very well. Wilkes College had a dance. I just can't remember all of them. . . .

There was a trial of some men who were mining too close to the river. I guess that's what it was all about. We had to go to the trial but nothing came of it. I testified before them. They just wanted to know if I was married to him and [that we had] four children. I think that the oldest two came to court with me. I don't think they wanted the two younger ones. Then after that I don't think they wanted [any of] the children in court, as I recall. But they weren't found guilty of what they had done. . . .

WOLENSKY: Did you have a memorial service for your husband?

STEFANIDES: Yes, we did have one. It was maybe a month later over at Holy Trinity Church. They put a casket in front of the altar. I don't know, maybe I just wanted to let go. It wasn't big; it was just for the family. The memorial service [for all the victims] was up there at Port Griffith, in St. Joseph's Church. This priest has a mass every year, every year. About twenty-five years ago they put that monument up. There were quite a few people in attendance [for the dedication]. Of course, there is a mistake in the monument. I told my oldest daughter, I said, "Look at the name on there." She did go and tell the ones who took care of it, but they just didn't seem to care. It's Donald Stefanides, instead of Daniel. It's on the bottom. . . .

## ORAL HISTORY: JOSEPH STEFANIDES

Joseph Stefanides (figure 42), a miner, had been working at the Knox Coal Company for fifteen years when the disaster occurred. Just two years earlier he had secured a job for his younger brother, Daniel, who labored in his crew. Joseph would almost certainly have been caught in the disaster if not for a head injury that sent him to the hospital in early January, 1959. He remembers learning about the crisis while still in the hospital. (From Joseph Stefanides, taped interview, September 15, 1990, NPOLHP)

JOSEPH STEFANIDES: I worked [for the Knox Coal Company] at what they called the Schooley Slope. When we were running out of coal there towards the end, they opened up in Port Griffith. We were the first ones to go up there, to the River Slope. Mike Lucas and myself and Billy Sinclair. . . . We had a chamber. See, we drove chambers out of that slope. We had four or five different chambers working at one time. We had a chamber on the left hand side of the slope. That's where I got hurt, on January 8, 1959. I was unconscious from the eighth to the twenty-first.

ROBERT WOLENSKY: What happened to you?

STEFANIDES: A fall of roof rock. Right on top of my head. . . . I'm lucky to be alive. You could put your fist in my head. [It happened] in that chamber off the River Slope.

WOLENSKY: The one that broke through?
STEFANIDES: Yeah, the one that broke through.
WOLENSKY: I see. And were you knocked out?
STEFANIDES: Yes. . . . [On January 22nd] Merle Ramage [who had escaped with Frank Handley] came in the bed next to me in the hospital, and I said, "Merle, what are you doing here?" He said, "I just came for some tests." I had found out the river broke in [that same afternoon] as I watched the news. I asked him, "Where's Danny?" He said he didn't know, so I knew he was lost. Then I went berserk. I did and how—because my brother was [in] there. They had to put me in a straight jacket. . . .

I tried to tell them. They were dumping railroad cars and what not trying to plug that. I was calling everybody including the mining inspectors and [telling them], "Don't try to plug it because you won't plug it, it's twenty-eight feet high in there. When big railroad cars hit it'll just eat them right up." I said, "You have to dam it." So after two weeks [two months] of hell to pay, they dammed the river—after they had it flooded. But if they dammed it in the first place they wouldn't have had this flooding. . . .

WOLENSKY: Your brother Danny worked with you?
STEFANIDES: Yeah, because he had a family to support and most of his business was on credit. So he says to me, "Joe, how about getting me a job?" He said, "I'm going broke here [gas station]. I can't make a living." I got him a job at the Schooley Shaft. I got Danny in. . . . [He was] a hell of a nice guy. We got along wonderful.
WOLENSKY: Where was Danny working when this happened?
STEFANIDES: He was with Mike Lucas and Billy Sinclair. I don't know whatever happened but my opinion is that when Mike Lucas got the word that the river broke in, he called them. See, all he would [have to] do is shake the chute a couple times and they'd come down [out of the chamber]. Whether he did that or not, I don't know. I wasn't there. Mike got word that the river was coming in. He got out. Why didn't the laborers get out?
WOLENSKY: Were the laborers at the face?
STEFANIDES: Yeah. Mike was topping cars. I wouldn't top a car. I figure the miner belongs within the face, not in the gangway.
WOLENSKY: So Mike sort of had a head start?

Fig. 42. Joseph Stefanides, miner, Knox Coal Company.

STEFANIDES: A good head start. Well, I excuse him because he has sort of a nervous condition, Mike does. I went down to see him, in fact. He said he didn't know what happened. He sent them his signal to come down, but he said he didn't know what happened.

WOLENSKY: How many years did you work together?

STEFANIDES: It was a few years. It wasn't too long because Danny got killed a few years after he was in the mines. He was thirty-three when he got killed.

WOLENSKY: Did you guys drink and pal around together?

STEFANIDES: Yeah, we use to stop for a drink every day at Jellies when we worked in Exeter. But after we worked in River Slope we went straight home. . . .

WOLENSKY: Did you have any idea where that [fatal] chamber was going?

STEFANIDES: Well, I told that surveyor. I said to Joe Stella, the surveyor, "Joe, we're under the river." I said, "We're gonna get hurt one of these days." He said, "No, we're along side of the river." And that was it. . . .

WOLENSKY: Were you worried where that chamber was going?

STEFANIDES: Sure. I saw the blueprints. I asked Joe Stella for them.

WOLENSKY: And you could see where it was going?

STEFANIDES: Right.

WOLENSKY: What did Joe say when you when you pointed it out to him?

STEFANIDES: He said the way they're mapped out they're not under the river, so. . . .

WOLENSKY: Was Billy Receski marking the chambers for you?

STEFANIDES: Yeah.

WOLENSKY: Did he have any worries about where you were going?

STEFANIDES: I don't know. I couldn't say that.

WOLENSKY: Who was in that crew when you were first driving the chamber?

STEFANIDES: Mike Lucas, myself, Billy Sinclair, and my brother Dan was with us.

WOLENSKY: That was one of the first chambers you drove then?

STEFANIDES: Yeah. As we were driving the slope, if we had problems we come back to this chamber for a day or two's work. And that's what they did before it caved in. They hit problems where they were going, so they came back to this chamber until it [the other work] was straightened out.

WOLENSKY: So this chamber was kind of a "keep yourself busy" chamber?

STEFANIDES: Right. . . .

WOLENSKY: Do you blame anybody for that disaster?

STEFANIDES: I don't blame anyone.

WOLENSKY: There were some really good miners at the Knox Coal Company, Joe. You were skilled, Frank Handley, Bob Groves. You just wonder how this could've happened with the skilled miners you had.

STEFANIDES: Will of God.
WOLENSKY: Just one of those things?
STEFANIDES: Uh-huh.

### Letter to the Editor: Henry J. Kaveliski, a Victim's Brother

(From "Brother of Trapped Miner Writes Letter from Chicago," Pittston *Sunday Dispatch,* February 15, 1959)

    I, Henry J. Kaveliski, brother of Dominick, one of the 12 missing miners, read your copy of the Sunday Dispatch dated Jan. 25 and do wish to thank you for coming out with the truth. I am referring to the editorial, "Disaster Strikes," and I am sure if some other folks bought a copy they would also learn the facts of what's happening out there in the coal mines.

    After all, I came from a family of nine and at one time I remember when my father came home from the mines hurt so bad that he could hardly walk and after a couple years he got hurt again. We were four boys and three girls, I was the youngest of the boys. I remember when my three brothers, Anthony, Stanley and Dominick, also my father, worked in the mines all at the same time. I was lucky. I only worked 53 days and wound up in the Pittston Hospital.

    My brother Dominick was born in Duryea and he at one time worked for Kehoe-Berge Coal Co. and liked his job there. Kehoe-Berge never worked so close to the river and I do not believe the Knox Coal Co. should have. Well, that's too late now to mention. Anyway, I would like to know what was under the ground in Pittston, or Port Griffith as the place is called, when my father worked in the mines for 40 years or more and my brother Tony for 35 years and Dominick for 25 years, and all the hundreds of other men, thousands in fact? How much more coal is under there to mine, or were the 12 men robbing pillars?

    I was there the week of Jan. 25. I got to Port Griffith at 8 a.m. on the 28th of January and they were building a form to pour concrete and I was told that it would take seven days to dry, then they could put in a pump. Well, the 12 men in the mines sure would like to know that. After being down there for six days why it would be another seven before pumping could start? If some of the truth were only told I am sure it would be much easier on the 12 men's families.

Henry J. Kaveliski, Chicago, Illinois

### LETTER TO THE EDITOR: CAROLINE BALOGA

John Baloga's wife, Caroline, wrote numerous letters to local newspapers expressing her deep sadness and anger over the loss of her husband (figure 43). John Kehoe Sr., a coal mine operator, founder and publisher of the Pittston *Sunday Dispatch,* and a powerful political figure in Pittston and Luzerne County, included the following letter, along with his reply, in his weekly column. (From "As Kehoe Knows It" Column, "Letter From Widow of a Knox Victim," *Sunday Dispatch,* February 22, 1959)

Following is an interesting letter I received from the widow of one of the men entombed in the Knox Coal Co. mines:

Dear Mr. Kehoe:
   Writing to let you know that I read your column in the *Sunday Dispatch* and sure enjoy reading it. You sure tell the truth so I'm asking you to give me some information.
   I am a widow of one of the entombed miners and people tell me to sue the company, but I hear Knox is broke. I live near the slope where the river broke in and I see they are cleaning up scrap such as iron and anything that brings money in. I did see a lawyer and he said he doesn't know if we can sue because we're getting compensation. I get $75 every two weeks and also Social Security, but for how long will that last? I have four children, three under 18 and one 21. He [oldest son] works two days a week and at times there's no work for two or three weeks and it's not enough for him for clothes. I haven't the heart to turn him out because I need him at home to try to take the place of his father, but he will never do, as my loved one was a good man who never drank and was good to my children and me. The longer he's gone, the more I cry. I ask God why He did this to me, because my husband didn't deserve a death like that. A flick of the finger and he was right there whenever I needed him and now I

Fig. 43. The Baloga family, photograph taken immediately after the disaster. From left to right: Audrey (age fifteen), John Jr. (age six), Donald (standing, age twenty-one), Sandra (age thirteen), Mrs. Caroline Baloga. (Courtesy of Audrey Baloga Calvey)

have to pay to get something done and I just can't afford it.

He left me with a lot of bills to pay and $1,000 on a car, so what I got from the disaster fund, I spent and I'm without funds and it isn't much for me to go on the checks with three children to take care of. Sometimes I just feel like going to meet my loved one because I'm so lonesome and despondent without him. I know I'm going to live a hard life without him because he brought me more than I'm getting now and I can't understand why it had to be me, because he was the best man God took. I don't go out at all. All I do is stay home and think about him from day to day.

I hear now that they're not going to look for the bodies. How can they do this to me and the other families? I'd like to give my loved one a nice burial and visit his grave. This I can't have. The state should have handed some of the money to us instead of putting all of it to seal the mines. It doesn't bring our loved ones home to us. The company would look for the bodies if the state paid, but it's up to the Knox and they won't waste all the millions to clear the mines just to look for the bodies.

Mr. Kehoe, I guess you felt bad when your loved ones were left in the mines years back.[1] I read about it in your column. I don't know what to do or where to go for information. I'm not too well and I'm always having a doctor because I'm too heavy on my feet so I have to stay off them as much as I can. I had my husband's due bills [pay receipts] over at Knox in Exeter and I just got them back as my lawyer wanted them. I see where they took some of the old ones out which have a bigger pay and marked new ones and put them in where the old ones were, but the Wage and Hour men were at my place and marked everything off the due bills before the Knox took them.

I don't know what my lawyer will do, so please let me know if I can sue. Even if I'm getting compensation they can't stop it and make me return all that I've received already. Please let me know. I wish the state would make them look for the bodies or make Mr. Dougherty help us out as it was done by him and the other greedy mongrels. And if Knox is broke can we sue Penna. Coal Co. or both? Don't disappoint me. I enjoy reading about what you have in the *Dispatch*. I was going to come up to talk to you but I was afraid I wouldn't be let in to see you. I passed your place at Harding going for a drive. Thank you and God bless you.

Yours truly,
Mrs. Caroline Baloga
12 Rose St., Port Griffith

### Reply from John Kehoe Sr. of the *Sunday Dispatch*

As to the above letter, I advise Mrs. Baloga to engage a good honest lawyer like Atty. Frank Flanagan who will not charge her unjust or excessive legal fees. Even if the Knox Coal Co. is broke, Bud Dougherty and Louie Fabrizio should not be released from their obligations to the distressed families of the Knox Disaster victims. They should be compelled to see that a search is conducted for the victims' bodies and if they refuse they should be jailed until they do it.

I realize first hand what a terrible thing it is to lose your loved ones in a mine disaster and to lose the family provider. I would like Mrs. Baloga to know that she shouldn't be fearful of not being allowed to visit me at my home in Harding. The gates at my home are always open to decent people like her and I'm always willing to talk to them.

## Letter to the Editor: Caroline Baloga

(From "Widow Deeply Hurt," Wilkes-Barre *Sunday Independent,* April 1959)

Please print this answer to the Retired Mine Foreman who said, "It would take millions…" I am the widow of one of the trapped men in the Knox Coal Co. mine. I read your article and was deeply hurt as you asked what are they going to find down in the mines.

You mentioned that Governor Lawrence is spending the taxpayers' money. Well you aren't the only one paying taxes. There are others in Pennsylvania who are paying taxes too. And it seems they aren't complaining about it.

I think the people are sympathizing with the 12 families saddened by our great loss and are hoping that workers can get into the mines real soon to find the 12 men and bring their bodies out.

Perhaps you have no heart or have no feeling or you would never write the letter.

What do they expect to find down there? You mentioned mine cars, gondolas, dirt and everything that was thrown in the hole. But did you forget there are 12 bodies still down there and what they went through before their lives were taken away? And they were taken away from us, their wives and children.

I advise you to say a prayer and thank God that you are safe and retired and still alive on this earth. I'm sure the 12 men would like to be safe and retired as you are, but there was no way to reach them in time.

So here I am alone after 20 years of married life thinking of the happy Easters we had together. But this one was the saddest of all. We just gulped our food and didn't care to even eat, because the one that meant the world to us wasn't at the table to share our Easter blessings with us.

I know he is in God's Heaven because he was a very good husband and father. And that's more than you can say, or your wife can say if you have a wife.

I've had people write to me and even send me checks because their heart and grief was with me in this awful lonely life I'm now living.

I thank each and everyone who was so kind in helping me out. It will never be forgotten and I pray for all who helped me and are still helping. They must know what it means to lose a loved one as I did.

It's been so long each day. I go out to the porch and watch what is going on at the place where the river broke in and see everything that goes on. So you see it's very hard for me to bear. He is so close and yet I can't go to see or talk to him, but I know he is watching over us at home, from God's Heaven.

God only knows how much longer I'll wait before they get into the mines and look for him. I hope and pray it's real soon, because there isn't a day goes by that I don't cry when I watch the people coming to see the place. So you can imagine what I'm taking into my heart.

So if I were you, I wouldn't worry about the taxpayers' money. I think God will reward them in some other way. I'm sure I wouldn't worry. I'd help all I could if it happened to someone else rather than my husband. But I'll hope and pray nothing like this ever happens to anyone, as it's sad and heartbreaking to bear.

Mrs. John Baloga, 12 Rose St., Port Griffith

### Letter to the Editor: Caroline Baloga (figure 44)

(From "Knox Miner's Widow Gives Her Views," *Times-Leader Evening News,* August 9, 1959)

. . . I am the wife of one of the 12 men who are still down in the Knox mines. With God's help I know he will come to me and my children, in what condition, I don't know. But I know he will be satisfied where he is going or is "gone."

All I do is sit, walk, and cry—perhaps a bite or two once in a

Fig. 44. John and Caroline Baloga, December 1958. (Courtesy of Audrey Baloga Calvey)

while. "How long will this last?" I ask. Weeks went by and perhaps months but I haven't heard anything much, just as you can't hear anything from the men who are at fault. Those men and other men who worked at the Knox knew that to make a living they have to keep their mouths shut just as my loved one did or he would have been out of work the next day.

The bosses knew when they had the "bull by the horns" because if the men opened their mouths they would be out of work and they couldn't get work at any factory if they were a certain age, and they couldn't get their unemployment checks because the bosses would say they quit. And to get a pension they are too young, so what could a poor miner or laborer do but do what he was told?

I know because my husband told me time after time that he worked in bad places and in wet places up to his hips [in water]. If there was no water in River Slope as Mr. Fabrizio said, why did all the men who worked there wear boots? My husband was one of them who came home wet and muddy many times.

And why would he tell me the pillars were as thin as kitchen walls and if the water broke in they would be all like dead rats and I believed him just as I believed in what Miss Zelonis said her trapped brother often told her. No one knows what I am taking into my heart because the only one we loved and will ever love is taken from us by "the money-grabbers." I don't feel sorry for myself as much as for my husband, the way he had to be taken away from me and my children.

Even a dog doesn't deserve what those 12 men got and I hope and pray that those at fault will never forget it because they stuck with the company and not the miners. I know because the accident happened earlier than the time they said it occurred. Maybe the men would be home today with their loved ones if they took action and notified the men as soon as it happened. It was said at the trial in Exeter that the water broke through at 12:40 or 12:45 p.m., but it isn't so, because it happened between 10 and 10:30 a.m., and if they notified the men to get out, that the river had broken in, they would have run instead of walked when the only order issued was "all men out."

I hope they make those responsible suffer as we are doing, all the 12 families and their children. You may print this any way you want to but I want them to know just how I feel.

Heart Broken Wife and Mother [Mrs. Caroline Baloga]

### Letter to the Editor: Caroline Baloga

(From "Disaster Fund," Wilkes-Barre *Sunday Independent,* August 22, 1959)

At the Knox disaster where 12 men aren't looked after, seven months are gone by during which they didn't even try to find the ones we love.

A disaster fund was raised and Mrs. Min Matheson, whom I highly praise, came to see me and said, "Here's some money."[2]

I thanked her and cried as she stood by my side. The ILGWU [International Ladies' Garment Workers' Union, which Mrs. Matheson headed], I thank a lot. What money they gave helped me a lot.

I wish the Knox was run by Min Matheson for she is good, brave, and strong and with her nothing would go wrong.

I thank you from the bottom of my heart.

Mrs. Caroline Baloga, 12 Rose St., Port Griffith

### Personal Letter: The Benedictine Fathers of Holy Trinity Monastery to Caroline Baloga

The Benedictine Fathers sent the following reply to a letter sent by Mrs. Caroline Baloga. (From *Times-Leader Evening News,* December 8, 1959)

Dear Mrs. Baloga,

Thank you for your letter. God has given you a tremendous cross to bear in life, but it is good to see that you have such faith and courage. God is never lacking in supplying us with the graces we need. For some very important reason God wished to have your husband at that time of that disaster. We cannot understand these things. Maybe your husband was more prepared to die at that time than he ever was or ever would have been. And maybe he is now in a better position to help you in some unforeseen way in your life in time to come. Only God knows these things. His ways are difficult to understand at times, very difficult.

It is hard for the heart to bear the anxiety about not having given a loved one a decent burial. Yet there is always the consolation that one day we will all rise, whole and entire, from our graves, to meet our Creator. The fact that your husband lies somewhere beneath the earth will make me think more of his departure than if he were already laid in a regular formal grave. And the only thing that will help him most now is prayer, prayer and lots of prayer.

The greatest thing we can do for you at this time is to comfort you with our prayers and to beg God to bless you in a most special way—you, your family, and all others who were involved in that terrible disaster. We pray every day for all our friends and benefactors. You will be in those prayers, too. Continue to have faith and patience. God will see things through for you. And people are good, too, as you mention. May the peace of Christ be yours at Christmas time.

The Benedictine Fathers

### Letter to the Editor: Ervin Featherman, a Victim's Brother

(From "Wants to Know," Wilkes-Barre *Sunday Independent*, July 5, 1959)

As brother of the late Charles Featherman who was one of 12 who met death in the Knox mine disaster, I am naturally keenly interested in that tragedy.

I should like an explanation of why the company is preparing to seal off that part of the mine in which the men were trapped on January 22, instead of continuing to search for the 12 bodies. I should also like to know what disposition was made of the money provided by the state for the de-watering of the mine.

Ervin Featherman

## Letter to the Editor: Caroline Baloga Replies to Featherman

(From "Answers Featherman," Wilkes-Barre *Sunday Independent,* July 12, 1959)

I read the letter last week by Ervin Featherman who has a brother Charles entombed in the Knox mine in Port Griffith. He wants to know why they are sealing that part of the mine where the men are entombed. The mines are not getting sealed as yet. They are trying to seal the hole where the water broke in and are having a wall put in the mine around it. As for sealing the mine they may do it later on.

I too am concerned about the men because my husband is also in the mine. As for the money, I think they wasted most of it. The state is not here to see where the money goes and I think they are getting as much as they can because it's not the company who is paying. A lot of men are taking their time and taking the money as long as it comes so easily. They didn't even start looking for the men and I don't think they will. They should have shared some of the money with the families. They can put it to good use and not waste it like it is wasted here. I live near the place and I see what is going on.

The ones at fault didn't even come to help us out as others did. Why don't the families start suit against the ones who were looking out for only themselves? They didn't care for the men as they don't care to look for them now. I wish you or your brother's wife would come to see me or write me as I watch from day to day what is going on. They just don't care about the men or their families. We just have to live with it for the rest of our lives and it's very hard to take. But with God's help I think we can carry on. We didn't do wrong. But the ones who did will never rest right. I hope the memory never leaves them, for what they did to our families.

Mrs. Caroline Baloga, 122 Rose St., Port Griffith

## Oral History: Ida Gizenski, Joseph Gizenski Jr., and Alfred Gizenski

Ida Gizenski lost her husband, Joseph Sr., and Alfred Gizenski and Joseph Gizenski Jr. lost their father in the Knox Mine Disaster. Mrs. Gizenski was thirty-four years old at the time, and her sons were seven-

Fig. 45. The Gizenski family. From left to right: Joseph Gizenski Jr., Ida Gizenski (holding a photograph of her husband, Joseph "Tiny" Gizenski), and Alfred Gizenski.

teen and fourteen, respectively (figure 45). They reflect on the disaster and the loss of their loved one, a tall and muscular miner known as "Tiny." Mrs. Gizenski also describes a strange occurrence the night before the tragedy and the difficulty securing her widow's black lung pension. (From their taped interview, July 10, 1997, NPOLHP)

JOSEPH GIZENSKI JR.: I was coming home from wrestling practice. It was seven o'clock. I heard about it on the car radio. We knew right away. Knox Coal Company in Pittston, I knew that was where my dad worked but I had no idea that my dad was involved, really. That was his last day of work there. They were finishing up and at noon they would have been done. I think this broke approximately 11:35 a.m. See, dad was a rock contractor and he drove the tunnel to the coal. When one job was done he went somewhere else to another job. The contractor was Stuart Creasing. Knox Coal Company issued the contract.

IDA GIZENSKI: Well, nothing happened that same day. We just sat and waited and listened to television. About twelve o'clock at night two men came down. I don't know who they were. All I know is two men came and asked if I had known this happened and that he [my husband] was one of them? But they didn't tell me he didn't get out.

ALFRED GIZENSKI: I drove up [to the disaster site], me, one of my buddies, and my Uncle John. He would be my dad's brother. [We saw] mass confusion, really. I couldn't see much because it was dark, very dark. You could hear the river's whoosh swirling as it was going down into the mine. That was a loud sound. They had started to drop mine cars [into the hole] at that time, and then later they brought the big rail cars and started dropping [them] in. We must have been there for about three hours. We were told by somebody that there was no sense in staying there. There was nothing else that could be done right at the moment.

Cold. It was cold. A lot of ice in the river. Freezing. We were down below, right where the gap [whirlpool] was, right in that area. My dad's pickup truck was parked near there so we went to the truck first to see if it was unlocked or if his clothes were inside. Maybe he got out, you know, but . . . [not so]. We brought the truck home that same night. I went up about the third or fourth day to get dad's clothes. He wore street clothes and then changed into his mining clothes, boots and stuff because it was always wet. I got his clothes. . . .

ROBERT WOLENSKY: How old were you when it happened, Al?

A. GIZENSKI: Seventeen. I was a senior. I was graduating in June.

WOLENSKY: Joe, how old were you at the time?

J. GIZENSKI: Fourteen.

WOLENSKY: Were you at school when you heard about it?

J. GIZENSKI.: Well, I don't remember, really. I think when I got home, that's when I found out about it. . . .

WOLENSKY: What did you do over the next couple of days?

A. GIZENSKI: The next morning my *dziadek,* that's grandfather Gizenski, he wanted to go up, so I took him up. By that time we were able to walk down by the river because they had already filled in a lot of dirt. But the water was still going in the hole. They were still contemplating whether they should put any more cars in to try and plug it up. It was noon because I remember we stopped and got a sandwich on the way back down so it was just about noon when we left there.

At that time we knew that there were twelve men that didn't get out. We hadn't seen or heard from dad so we were just kind of contemplating that he was one of the twelve. He would've been working probably about sixty feet below where the river broke in. See, we had an idea of this because he used to tell us about where he was working. He also used to talk about how he could hear water. He talked about this for about two weeks because this was a time when the river was up. It was up, up, up and then it froze, and then it thawed, and that's when all this ice broke loose. He talked about [it] all the time. He said, "We could hear water. We are under the river." That's one thing that wasn't supposed to be because there's a red line no matter where there's a mine, there's a red line when it comes to water. He told us that one of these days that river is going to break in.

I. GIZENSKI: Yes, he'd say that when he would be talking with his buddies when they'd come to the house to visit. He'd tell them that they were too close. . . .

A. GIZENSKI: Well, see, dad kind of knew that. He kept saying, "We're under the river and we're not supposed to be. There's a red line and I think we're over the red line." But he had no idea where the red line was. He mentioned that several times. . . .

Dad was a twenty-four-hour a day worker. He didn't only work in the mines. He worked for himself. He dug wells. Laid them up with stone, poured concrete, hauled coal, hauled sand, laid block. [He was] a big strong man. . . . . If there was any way to get out of that mine he would have been out of there. But there was no way. They had nowhere to go first of all. You envision a tunnel coming to a dead end, and the river breaking in behind you. Where are you going? And it was sixty to one hundred feet, maximum. What can be done? So they were the first persons to drown.

J. GIZENSKI: Those three men were the first three men to drown.

A. GIZENSKI: It was quickly. They probably didn't know what hit them. They were shut in. That's the way I was told by the other miners that I talked to when we went up for his clothes, especially one guy that dad used to get wine from—dad used to get a gallon of white and red wine from him. All they had left to bring out was their tools and themselves. He said they were that close to being done. . . .

WOLENSKY: Mrs. Gizenski, you mentioned on the phone that there was a strange occurrence the night before the disaster?

I. GIZENSKI: Well, [we heard] a loud noise on the living room ceiling when we were watching TV. It sounded just like somebody dropped something heavy. I said to him, "What was that?" He passed it off as maybe a rat was up there [attic] and knocked something over. But I didn't think so. I thought it was a sign of something. I coaxed him not to go to work the next morning because it was such bad weather. But he said, "No, we only got a couple hours' work and we'll be moving out. We'll go get it finished." And he went.

WOLENSKY: Did you go up to the attic to check what had happened?

I. GIZENSKI: [I did check but] nothing fell. Just a big thump, and loud!

WOLENSKY: You thought it was a sign [that he shouldn't go to work]?

I. GIZENSKI: That's what I thought. I never said anything to him because he didn't believe in this, but I did. . . .

WOLENSKY: How did you get by after the disaster?

A. GIZENSKI: How did we get by? Well, it wasn't easy. People told us we were entitled to Social Security, stuff like that. Of course, it took a while to get this straightened out. Most everything was done on our own. We got a lawyer and some of these matters he brought to our attention. But as far as the people from the coal company. . . [nothing]. Well, yes, we did get a food basket from the coal company and then we also got a few baskets from two newspapers. A lot of our friends and neighbors tried to keep us up as well as they could. After a while, I mean things started to fall in place a little bit, but as far as the company themselves, or people from the company, not much of anything. They made themselves scarce, and I understand why now, but I couldn't then.

WOLENSKY: Why do you think they were scarce?

A. GIZENSKI: Because they knew we were wronged. In fact that Louie Fabrizio was convicted. He was one of the owners.

WOLENSKY: You finally got a settlement after how many years?

I. GIZENSKI: Seven. Seven years before they settled.

A. GIZENSKI: I remember we had about a $10,000 settlement from the suit and he [lawyer] took 33 1/3 percent of that. To us [our dad] was our livelihood, you know. No one else. I was in school and I didn't work outside [of the home] but I helped him because he also did a lot of well drilling. That's fifteen, sixteen, seventeen years old I was running that machine while he was at work or going to do something else. He'd come back and check on me. He was the total bringer of the pay, you know. I mean, he kept us in groceries and housing and clothing.

WOLENSKY: So you had an economic shock as well as an emotional shock?

A. GIZENSKI: Definitely.

WOLENSKY: Mrs. Gizenski, you mentioned earlier that you experienced serious emotional shock afterwards.

I. GIZENSKI: Yes. I was here but I wasn't here. I was far away (laughs). I was in the hospital for what, about three weeks, wasn't it? But I finally got straightened out.

A. GIZENSKI: It was a good year or year and a half until she finally got straightened out the first time. . . .

WOLENSKY: Did you get black lung pension?

I. GIZENSKI: Twenty-five years it took. My husband's relations wanted us to take a lawyer from out of state. But we didn't know anything about lawyers or anything at this time, so we just took one from here. . . .

WOLENSKY: Did rock contractors have a more serious type of asthma?

A. GIZENSKI: Rock dust. It's heavier; it's supposed to be worse [than coal dust]. I can't say. All I know, dad was thirty-seven years old, and at thirty-seven years old, he couldn't run too far.

WOLENSKY: Why did it take twenty-five years to get the black lung pension?

I. GIZENSKI: They kept denying me all the time.

A. GIZENSKI: They kept denying us because they claimed there was no proof that he had [it]. See dad didn't want to go to the doctor much. He never had X-rays, never had much of anything. I mean except for an injury that he got in the mines, hurt his foot one time, put a pick through his boot and his foot, he never missed a day's work no matter if it was in the mines or whatever he was doing. And you know we talked to different people and ironically how we got black lung was an accident. She talked to, who was that, one of the caseworkers?

I. GIZENSKI: Down at the [Social] Security office.

A. GIZENSKI: One of the caseworkers at the [Social] Security office, he asked her, "Are you getting black lung?" She said, "No, I've been denied four, five times." "Well I can't understand why." The caseworker got into it. He was a guy that had interest.

I. GIZENSKI: It was a lady.

A. GIZENSKI: She had interest in it because they figured she deserved it. So she checked into it. Well, we were all grown [by then]. She got what she deserved anyhow. I mean she deserved it from the word go....

WOLENSKY: Mrs. Gizenski, how did you feel about not recovering your husband's body?

I. GIZENSKI: Well, it bothered me for a while. But I knew I would have to live without him, so I just made the best of everything.

A. GIZENSKI: It was something we've never gone through. When your grandparents die you can go to a cemetery and there's the headstones, but that doesn't bring your grandparents back to you. In essence, we don't know where dad is at. I think we do because I don't think he had any way out of there. I don't think the river carried him anywhere. I don't think [so] because he would have had to go the long way back up hill to go anywhere else. From where that river broke in, bodies could have only floated that far. I don't think they ever got out of the water, so I think their burial place is where they ended working. All three of them. I think they're all three, right there. Right at the head [face]. They were washed right up against the head.

## NOTES FOR CHAPTER FOUR

1. Mrs. Baloga is referring to the Twin Shaft Disaster of 1896, where seventy-five workers died in a mine cave-in, including Kehoe's father, brother, and uncle.

2. Min Matheson served as the head of the Wyoming Valley District of the International Ladies' Garment Workers' Union between 1944 and 1962. On her work, that of her husband William Matheson, and others in building and maintaining the union see Kenneth C. Wolensky, Nicole H. Wolensky, and Robert P. Wolensky, *Fighting For the Union Label: The Women's Garment Industry and the ILGWU in Pennsylvania,* (University Park: Penn State University Press, 2002). See figure 61 on page 202 for a photograph of Min Matheson and another ILGWU member, Clem Lyons, as they were delivering goods to the Baloga family.

# CHAPTER FIVE
## THE CHILDREN

*I don't think they should call it the Knox Mine Disaster. I think they should call it the Knox Mine Murders because I feel as though everybody was paid off, everything was under the table and hush-hush, and double-dealings.*
Anita (Ostrowski) Ogin, daughter of victim Eugene Ostrowski Sr.

*. . . if that happened today, all those people would be put in jail. All those people. I mean, you wouldn't have had a mock trial down in a basement of a church because people would be [really upset]. It's sickening; it's just sickening. And it's not only because my dad was involved.*
Linda (Boyar) Davenport, daughter of victim Benjamin Boyar

Most of the victims' offspring were young and lived at home, although some were older and had moved away. The following letters and oral histories convey the children's sense of loss, sorrow, and (typically) anger over the death of their fathers.

## Letter to the Editor: Theresa (Altieri) Aiello, a Victim's Daughter

(From "Missing Miner's Daughter Writes of Knox Disaster," *Times-Leader Evening News,* February 1959)

Writing as a grief-stricken daughter of one of the entombed men, I feel compelled to criticize the laxity of everyone connected with the supposed rescue operations at the Knox Mine.

I feel justified in saying the entire community and surrounding areas are appalled at the apparent unconcern for the entombed men and their families who carry the brunt of this tragedy and who have not had one word of consolation from those directly responsible. Had it not been for the goodness of friends and neighbors plus the outstanding work of the Salvation Army and others too numerous to mention, these families would have gone unnoticed and un-cared for.

Perhaps if this tragedy had occurred in another state or [if] the men were of some importance in some political form, the fate of each one would have had some significance. As it stands, all hope for the survival of these men has faded and someone should be made to pay for it. Their lives were just as important as others no matter what position in life they held.

It's about time the people of this community stand up and speak their piece. I think this tragedy has opened up their eyes to just what they are up against. Let's not give those concerned another opportunity to entomb others as they have these 12 poor souls. Let the two and one-half million-dollar appropriation from the government be put to good use for the families of the entombed and the community, not as a mausoleum for God-fearing men.

Mrs. Theresa (Altieri) Aiello
440 East 118th St.
Cleveland, Ohio

Fig. 46. Members of the Altieri family. From left to right: Yolanda Altieri Parente, Ann Altieri Ferrare, and Frank Ferrare, daughters and son-in-law of disaster victim Sam Altieri.

## ORAL HISTORY: ANN FERRARE, YOLANDA PARENTE, AND FRANK FERRARE

Ann Ferrare and Yolanda Parente lost their father, Sam Altieri. Frank Ferrare is married to Ann. Together they discuss the disaster, its aftermath, and their lingering anger toward the company owners (figure 46 and 47). (From Ann Ferrare, Yolanda Parente, and Frank Ferrare, taped interview, June 21, 1992, NPOLHP)

ANN FERRARE: What I remember about the Knox mine disaster [is that] I was coming home from work on the bus with my neighbor and she asked me if I heard about this disaster down at the Knox mine. I said, "No, I didn't hear anything." Then it just hit me like a ton of bricks. My father works there! So I got off the bus and I rushed home and called my mother who was working at the time. I thought, well maybe she's home, so I called and—no answer. She wasn't home from work yet. So I called a next door neighbor because she knew when my father came home. I said, "Did my father get off the 3:30

Fig. 47. The Altieri Family, photo taken immediately after the disaster. From left to right: Mrs. Herman Ciampi (daughter), Sam Jr. and Vincent (sons), Mrs. Mary Altieri, and Mrs. Nicholas Foglia (sister). (Courtesy of the *Times Leader*)

p.m. bus?" She said, "No, I didn't see anybody getting off." I explained to her what happened. Then I called [my husband] Frank at work and I said, "I don't know, maybe my father's down there."

In the meantime my mother had come home. I called her and said, "I'm coming right down as soon as Frank picks me up and we're going to the Knox mine." When we got there we didn't know whether he was a survivor or what. I think it was the coldest day that we ever had in our lives. My aunts and uncles and cousins, we all went down. They said some survivors had just gone up to the Pittston Hospital. Well, we ran from there, we climbed the hills, to the Pittston Hospital. My father wasn't there. So then I guess we just knew that he was one of the victims. It was terrible. It was devastating. It was terrible.

It had to be about 4:00 p.m. And this happened at 11:00 a.m. but I don't think they tried to contact anybody in the family. They didn't contact anyone and if my neighbor didn't tell me I probably wouldn't have known about it until I turned the TV on. I mean it was terrible. Words can't explain it. Words just can't explain it.

So we came home and then, of course, we contacted everybody. My brother was in Oklahoma, Yolanda was in Cuba, my other brother, Frank Samuel, the youngest, was in school, Vincent was in Massachusetts, and then I had two half sisters from my father's previous marriage. His wife passed away and then he married my mother. One was living in West Pittston and one was in Ohio. Of course, everybody came home. I just can't explain how bad it was. A terrible, terrible feeling.

ROBERT WOLENSKY: When did you know that he would not come out?

FRANK FERRARE: When we got to the hospital we started pushing [so] they let us in the ward where all the miners were being treated. I started going from one bed to the other and asking all the miners about Sam Altieri. Nobody knew his whereabouts or they didn't want to tell us. But in questioning everyone I learned that one of the foremen found out that the Susquehanna River caved in and he called down into the mines to warn all the miners to get out. Sam Altieri was the first one to answer the phone. He told him to get all the miners out of the mines. So he goes on to warn all the miners and I guess the rush of water got ahold of him and just swept him away. If he wasn't a hero and warned all these other guys he probably would have been safe too, because he was right there at the elevator. I guess he was going from one vein to the other warning everybody. By that time the water came in so fast he just never had a chance to get out.

YOLANDA PARENTE: Well, I wasn't home when this tragedy happened. I was in Cuba married at the time, and my husband was stationed in Guantanamo Bay. The Red Cross came to the door and they said there was an accident. Well, for the Red Cross to come, I figured it was a pretty big accident. We didn't know the extent of it until our travels home. We saw it on TV in Virginia. When I saw it I knew there could be nobody saved in that accident. It was just terrible. When I arrived home everybody was here: newspaper people, neighbors, family, everyone, and it went on for a month. They kept telling us, "We're gonna get them tomorrow, maybe next week." It was a whole month they let us go on, people coming and going, before we knew there was no hope.

A. FERRARE: We kept holding on to this hope.

PARENTE: It was people from the mines, it wasn't Fabrizio himself. I just don't remember the men by name but they would show up every now and again at the door saying, "Oh well, you know, tomorrow will be the day; we'll get them tomorrow." There was never tomorrow. Then it was a week, and the week never came, and then it was a month. It was a month until they told us that was it.

F. FERRARE: The one fellow she was talking about was a union official named Frank. I forget his last name but he was one of the union officials and he kept in contact. They were gonna do everything they could to get these miners out but down deep I guess they knew that it was a lost cause because they were dumping railroad cars and they were dumping everything to try to plug this hole up. There was so much debris that they knew they would never get these miners out.

WOLENSKY: When they said they would get them out, would they get them out alive?

PARENTE: No, no just the bodies. Just get the bodies out, but it was just fruitless, I guess. They kept telling us these things and we just kept hanging on to hope.

WOLENSKY: Ann, did your father ever say anything about the working conditions at the Knox?

A. FERRARE: He did mention it to my mother. My mother did say that he kept saying, "One of these days the walls are gonna come tumbling in." He did mention it to her several times.

PARENTE: He worked with hip boots on. There was water there in the mines and this is how he went to work and, yes, they did complain and they were told if they didn't like the conditions go and get another job. My father was sixty-two; where was he gonna get another job?

A. FERRARE: He was gonna retire in February [1959]. He would have been sixty-two in February. Of course, the owners like Fabrizio, it took a while for him to come to the house to express his sympathy. I guess he was afraid to come alone so he had to come with our pastor. He came with our pastor [because] I guess he thought from reading the newspaper that everyone was so bitter. The families were so bitter and angry, they were so angry at what happened because it could have been avoided. It was weeks later. It wasn't immediately.

I guess we wanted to kill him. I mean, that was our father! For someone to walk in the house and you know that they deliberately took your loved one's life, because they knew what they were doing. They definitely knew what they were doing. They were just robbing the coal mines, they were robbing everything down there and they were using my father and all the miners as pawns really. That's why he came with our pastor. Our pastor tried to pacify us to soften the blow because he knew how bitter we were against him [Fabrizio]. Him and all the other cronies. He walked in and expressed his sympathy and we really didn't say anything to him. I think if we did start it would have been a scene so we didn't say too much and he just left.

PARENTE: I was here. It would have been a different situation if we didn't know the working conditions of these men, but we did know, and for him to come in and express his sympathy, that was really punching below the belt.

A. FERRARE: I think he did it mostly for the press to say that . . . .

PARENTE: . . . he came around.

F. FERRARE: He told us that where the river caved in the men were working with eighteen inches of rock cover. Right out under the river.

PARENTE: In talking about mine inspectors, they were just paid off. They didn't even go in to see the mines. They were met at the top of the mines, paid off, and that was it. Whatever was down beneath they never even knew.

WOLENSKY: Did you ever hear the men talk about that? How do you know that?

PARENTE: Well, because I've heard it. I've heard the older men talk about it. You know at one time there were [miners'] homes around here and there were five bars down in our little area, and all these miners would get off the bus, go into the bar and have a shot and a beer, and they would sit around talking about the day's events. My dad and my grandfather used to make their own wine. They had tons of friends, whether it was for the wine or just their personality, but we always had company. Our backyard was filled with the men all the time. The wine was out, the pepperoni, the provolone, the bread, everything. They would stay for hours talking about the day's events, what happened, and what happened years ago. You remember these things as kids. You just remember them.

A. FERRARE: And they never had anything good to say about working in the mines. It was just a way of earning a living.

PARENTE: Surviving, that's all it was. . . .

WOLENSKY: Did you sue?

A. FERRARE: My mother sued along with some others and all she got was $10,000 minus 33 percent for the lawyer's fees. So you figure it out. And she had a son in school to raise. Our church would give a donation and I think the ILGWU did too because I worked in the garment industry, and you would get $25 and $10 from this one or that one, but nothing really major.

PARENTE: The Jewish Community Center in Wilkes-Barre brought something.

A. FERRARE: Yes, we did we get food. Well that's a traditional thing when anyone dies in the area. The neighbors would always come over with food and they did come over with money, like someone would give us $10 or $5 or $20 or whatever.

PARENTE: This went on for a month. It wasn't like somebody died and you went to the viewing, there was a funeral and . .

A. FERRARE: . . . at least it was over. . . .

PARENTE: This was a month. Some people came like I remember the woman from the Jewish Community Center coming with a basket of food and that was probably two weeks after the event, so it was just like an ongoing thing.

WOLENSKY: Did that make it worse?

PARENTE: I don't think so, no. I think it was a good thing because people were kind and generous. They had sympathy for you and I think it was a good thing. That's what this area is all about. When anything happens like this, any tragedy, everybody pulls together.

F. FERRARE: You always had that hope that they were gonna bring the bodies out of the mines and for about a whole month this went on but down deep they knew that they would never recover the bodies.

PARENTE: You know there was never a wake or funeral or a grave you could go to.

F. FERRARE: This is the reason why this monument was set up. Sam DeAlba and some other people were very instrumental in setting that up.

A. FERRARE: He [Sam DeAlba] had no relatives [in the disaster]. He just took an interest in it.

F. FERRARE: He took the bull by the horns and got the community involved. In fact, they set up a fund now that is earning interest to maintain the monument, which is in front of St. Joseph's Church.

There are flowers there for Memorial Day. But he ran this from the ground up and he deserves a lot of credit for what he did. . . .

WOLENSKY: Some families reported some interesting things that happened, strange things, even premonitions. I wonder if you had anything happen?

PARENTE: I did. Yes, I told you in the beginning I lived In Cuba. The night before I dreamt that my father was dead and I saw him in the coffin. It was just funny because he always wore glasses and he didn't have his glasses on but I saw that as clear as anything. That was the night before.

WOLENSKY: So when the Red Cross came did you . . .

PARENTE: I knew.

### ORAL HISTORY: LINDA (BOYAR) DAVENPORT AND RICHARD BOYAR

Linda (Boyar) Davenport and Richard Boyar, children of Benjamin Boyar, were eight and thirteen years old, respectively, when the disaster occurred (figure 48 and 49). They reflect on the loss of their father and the meaning of the disaster to their personal and family lives. Like most victims' family members, they express a good deal of anger toward the company. (From Linda Davenport and Richard Boyar, taped interview, March 16, 1999, NPOLHP)

LINDA DAVENPORT: Well, I was only eight years old when it happened. My memories are vivid but yet vague. I remember coming home from school that day [and seeing] a lot of hustle and bustle at my house. People that weren't usually at my house were there. I remember asking my mom what happened. I remember my mother telling me to go to my half-sister, who lived across the street, on Wesley Street in Forty Fort. I was really kind of like removed from it all. My mother and my relatives were, of course, upset. I really spent most of that evening with my sister.

But I remember seeing it on the TV news that night and asking my sister [about it]. They kept putting me off and saying, "It'll be okay, it'll be okay." Then, as the days went on, my father never came home and it was just accepted. I knew that there was an accident at the mine and, it was just—he was never coming home again. That was it.

I was in the third grade. I went to school that day, but then I was off for I don't know how long. They just let me stay home. There were people from the newspaper that came to the house, our minister was there, and relatives from out of town came. There was a lot going on, a lot of things happening. But you just accepted that my dad was never coming back.

I remember that night, at my sister's house, even though she wasn't my father's real daughter [child from her mom's first marriage], she was very close to my father and I could feel and see how she was. I never saw my sister like that. Even at eight years old, I could see that she was upset and I guess I did cry.

RICHARD BOYAR: I want to go back to January twenty-first. I remember my father helping me with my homework that evening on the couch in the living room. We went to bed about ten o'clock and I

Fig. 48. Members of the Boyar Family. Linda Boyar Davenport (left) and Richard Boyar (right), children of disaster victim Benjamin Boyar.

Fig. 49. Dorothy and Benjamin Boyar, circa 1944. (Courtesy of Boyar family)

didn't even get to sleep yet, and I remember the doorbell ringing. It was about eleven o'clock and, sneaking downstairs, looking down the steps, I could see and overhear two men come to the door. One was telling my mother and father about problems they were having at the Knox mine down in the bottom veins. They were having trouble having the water pumped out. Something was wrong with the pumps.

They said that the water kept on building and getting higher and higher and the pumps couldn't handle it. They asked my father if he would mind coming to take a look. So my father, being an electrician working for the Pennsylvania Coal Company at the time and doing work for the Knox Company, he said, "Yeah, I will go down and take a look at it.'" He left on the twenty-first at 11:00 p.m. and that was the last time that I saw my father.

The next morning I awoke to a house full of people and I remember coming down the steps and the first one to approach me was Reverend Singer from United Methodist Church in Forty Fort. He said to me, "Looks like you are the man of the family now," and [he] shook my hand. Not knowing what he was talking about, I said, "What happened?" He proceeded to tell me that there was an accident at the Knox and that the river had broken through and my dad was trapped in the mine.

That day I remember a lot of people coming with food and so forth. My sister was in school. I didn't go to school because I had a cold. Later in the day someone took me up to Port Griffith and I remember standing on the riverbank right where the railroad tracks were [cut] towards the river, watching the water swirl around. I remember them throwing railroad cars down, bails of hay, anything to try to block the hole. I remember hoping for a month after that, thinking, well, there was always hope.

DAVENPORT: Because it was never really established that he was dead. My mother never really sat down with us and said, "Well, he's gone."

BOYAR: Never, no.

DAVENPORT: I mean it. You know, there was always a hope.

BOYAR: For a very long time, well longer than a week, maybe a month, it was like, they could still be alive.

DAVENPORT: I don't know if people were telling mom this or if we just had it in our minds that we weren't accepting it. I don't know, I don't remember. I know I felt like that.

BOYAR: Well, one thing I remember about the house and the days following—one person who really, really helped [us] cope with the situation and helped as much as she could was Min Matheson [of the ILGWU]. With money, with food, with all the help. I mean, she was on it. Anything we wanted or needed, she was there for us.

DAVENPORT: She took everybody in.

BOYAR: Yeah, she took everybody in. I remember her coming to our house, her being a woman of large stature. I remember her helping us with anything we wanted. You know, she was there. Min Matheson, yeah.

ROBERT WOLENSKY: Was your mom a garment worker?

BOYAR: No. My mother never worked.

DAVENPORT: We were better off than most of those families. We did live quite nice. Because [working in the mines] was not the sole profession of my father [who] had his own business on the side. I guess he made pretty good [money] from the Pennsylvania Coal Company also. And he was well insured, so our lifestyle did not change. She [mother] never had to go to work.

BOYAR: We still traveled, we still ate good.

DAVENPORT: I remember her saying she felt bad for those women who had small children and who didn't have it as well as we did. In fact, I think that was in the newspaper.

WOLENSKY: You said your father was doing work for the Knox.

BOYAR: He didn't work directly for the Knox. He worked for the Pennsylvania where he was a foreman. Many times, I went down in the mines with him, many times, on a Saturday when I was off from school and so forth. But, he wasn't a miner.

DAVENPORT: Although he knew the mines like a miner.

BOYAR: Oh, yes.

DAVENPORT: And that's what they kept saying to us. I remember when a state inspector came to the house and told my mother, "He knows the mine so well that he'll find a spot, he'll find a spot with little air pockets." But evidently, wherever he was, was not a good spot.

BOYAR: I was thirteen and Linda was eight. I remember hearing a story. I don't know how true this is but [it's about] Frank Burns from West Pittston, who was one of the twelve, but not a miner. He was over at the Knox at this time and someone told me the story that, as my dad was going to go down in the cage, Burns said: "Ben, where are you going?" Dad said, "I have to go down to check the pumps because there's something wrong down there." Frank said, "Well, listen, I'm not doing anything, why don't I just take a ride down with you?" He had no business in the world being down there. They were right at the bottom. And that was it.

DAVENPORT: I was very close to my father. My mother wanted to have children. She was previously married. She had my sister and she wanted to have more children. My father didn't want children, that was not his thing. She did talk him into having one and it was my brother. They were happy and my mother wanted to have one more. Thank God they did have a girl because this is what he wanted. He was a very, very good father, a good husband, and a good provider. I mean he was just the best.

BOYAR: Didn't drink, didn't smoke.

DAVENPORT: Well, a social drinker.

BOYAR: A social drinker, but I mean not go out with the boys and such.

DAVENPORT: Very health oriented for back in those days. Ate wheat germ in the morning, which back in those days, I don't even know where he got it! Today he would be in his nineties. You know, something makes me think that he'd still be here because he took very good care of himself. . . . I can still see the lunch bucket and I remember down in the basement he had that hat with the light on it. I remember his clothes because he wore the coveralls. . . .

BOYAR: He was about five foot, eight inches. . . .

DAVENPORT: He came from a large family and he was very close to his parents. They were born in Poland. My grandmother came over first but she was married to someone else and she had two daughters. Her first husband, I believe, was killed in the mines. My grandfather was a school teacher in Warsaw and he came over and met my grandmother, married her, and they had seven children. My grandmother had nine children and seven of those were my grandfather's. My father was born in Duryea and they moved to Dupont.

BOYAR: Boyar is shortened from Boyarski. My father changed that in 1950.

DAVENPORT: Only because of the business that he owned on the side and because of the people within the area. [We were] living in Forty Fort [and] the background with Forty Fort, the people knowing that you were, you know, Polish, or Catholic, in those days it was not easy.

BOYAR: He was the leader of his family. Very strong. He wasn't the oldest, but he was the leader. He called all the shots back then!

DAVENPORT: It was funny because for a small man in stature, all his brothers were big men. . . . I remember my mother later on saying he was offered many jobs outside of the mines and outside of the area. He wouldn't [go]. I remember her saying that if only he would have [gone], because she would have moved. We were little. He was offered many jobs outside of the mines [but] he just wanted that mine. He loved that life. . . .

WOLENSKY: What happened when you finally went back to school in January 1959?
BOYAR: I don't know who it was, I think it was one teacher who said that Richard is back in school and I think we all feel the same, that we feel sorry for the loss of his father. I don't remember anything special, nobody said anything. They called me in the office to say sorry, you know.
DAVENPORT: Maybe people would ask me how my mother is doing or something. But as far as special treatment, I don't know. . . .
WOLENSKY: Do you often think about your father and the accident?
DAVENPORT: Some things trigger it.
BOYAR: Yeah, certain things [like] the river looking the way it does.
DAVENPORT: Yeah, the river does [that to] me. It has a big impact on me.
BOYAR: Being out in public, people say, "Rich, I saw the Knox [on TV]," and you think about it then. But just to sit there and think, oh boy.
DAVENPORT: Yeah, January, you know. But, forty years, I mean, forty years is forty years. We have so much other stuff going in our lives with kids and work. You know how it is.
BOYAR: I've come to terms with it. I have my own life. I have my own children. There are plenty other things that keep me busy, keep my mind off of it. I don't know about my mother. You know, how it affected her. . . .
BOYAR: Those guys got off, and if that happened today. . .
DAVENPORT: True, if that happened today, all those people would be put in jail. All those people. I mean, you wouldn't have had a mock trial down in a basement of a church because people would be [really upset]. It's sickening, it's just sickening. And it's not only because my dad was involved. I'd still feel the same way.

### Oral History: Francis Burns Jr.

Francis Burns Jr. (figure 50), son of victim Francis Burns Sr. (figure 51), heard the breaking news about the disaster on the radio as he traveled home from class at King's College in Wilkes-Barre. He and his family came to accept the death without the accompanying anger so characteristic of most other families. (From Francis Burns, taped interview, January 21, 1992, NPOLHP)

My dad worked for the Pennsylvania Coal Company, not the Knox Coal Company. At the time of the incident he was a coal inspector. My recollection of my dad was that he was a hard worker, very industrious. Most of his life he was involved in coal mining. He was a coal inspector, a mine foreman, and, years back, a miner. During the Second World War he worked in New Jersey. He was a great cigar smoker, loved his cigars.

Fig. 50. Francis Burns Jr., son of disaster victim Francis Burns Sr.

We lived in a half of a double block house with the five of us in the family. We were close. My father was a decent, religious man who tried to instill in the kids a respect for authority and also for religion. My brother and I were both altar boys, my sister was active in church affairs, and same with my mother. My fondest recollection of him is he would always take my brother and me to baseball games in the local area; for example, in the Eastern League when the Wilkes-Barre Barons and the Scranton Red Sox were prominent minor league baseball teams. That was something that we looked forward to. Plus basketball games at the old armory in Kingston. It was a shock when this happened, and certainly we all missed my dad.

There are certain days you'll never forget in your life: where were you when Kennedy was shot, where were you when Pearl Harbor was bombed? I was at King's College, a junior, and it was January 22, 1959. I finished my last class around 12:30 p.m. and was headed home to Pittston. I had the radio on and that's when I first heard about it. Sometimes you get a gut feeling, a gut reaction and it was just something that hit me. I felt my father was one the entombed miners. So I went home and asked my mother if she had heard anything. She didn't so I went down to the area in Jenkins Township where the disaster happened.

I'd known a lot of my father's friends, other miners throughout the years. They would either be up at our house or we would meet them at a baseball game or something. So I was able to get to this one shifting place [and ask them questions]. I'll never forget this day. It was, in my opinion, the coldest day of my life. It was freezing and the winds were whipping up in Jenkins Township. One of my father's friends told me that, "Yes, your dad's one of the entombed miners" and that he went down to the Red Ash Vein. The significance of that is the Red Ash Vein is the lowest vein in the mines. My dad and Mr. Boyar were away from the rest of the miners so there was absolutely no hope of his survival. I knew that at the time, if he was in the Red Ash Vein there was no method of egress.

I went over to the Susquehanna where you could see the cave-in and the water swirling in. I stayed for a couple hours. I can remember coming home and I thought I had frostbite. In the meantime, we called my sister who was at home, and we notified my brother's fam-

Fig. 51. Mr. and Mrs. Francis Burns Sr., wedding photograph. (Courtesy of Burns family)

ily in North Wales, Pennsylvania. There was just nothing you could do.

My mother was very distraught. One sad part about it was that there was always hope, at least hope in her mind. I think myself, my brother, and my sister realized that there was no hope, but my mother always clung to that spark that something would happen. Naturally the television and radio announcers would say what steps were being taken to rescue the miners, and I think that kept a spark of a promise in her, which turned out to be fruitless. Later on in the evening I went up to the Pittston Hospital, which was located a matter of blocks from where we lived. But when I was told that my father was in the Red Ash Vein I knew that there was absolutely no hope for him. . . .

I believe my mother finally resolved in her own mind at that time, approximately a week later, that there was no hope and my father was just entombed. It was mainly grief and a period of reconciliation with friends and family. Later it became evident that possibly some illegal mining and some various other activities [had taken place]. It conjured up some grief again and also a little bit of anger but we resolved that it was just God's will and made the best of it.

Some of the miners who worked for the Knox Coal Company were able to bring legal proceedings against the Pennsylvania Coal Company. We could not because my father worked for the Pennsylvania Coal Company and my mother was limited to receiving workmen's compensation. I'm sure perhaps she could have sued the Knox Coal Company but they were bankrupt and it would have been an exercise in futility. And we weren't anxious to drag my mother through a lawsuit that in the end would mean nothing.

My mother got over it. My mother was a very religious person. She accepted it and went on with her life. Naturally every time around this time of year [January], with the publicity, it brings up the incident, but we've all gotten on with our lives. It was a bad episode, it's over, and we made the best of it.

Naturally the lasting impact was that my father was entombed in the mines and, from a personal viewpoint, I think it brought the family closer together. It gave us an understanding of togetherness, and I also think it gave us a closer, religious binding. We must accept disaster, we must accept things that we have no control over and continue to do the best we can. . . .

One thing I do have to tell you. I said I was a junior at King's College when this happened and after the incident at Knox, King's College, through [President] Father Kilburn, granted me free tuition for the balance of the year and for my senior year. It was a great act of charity, something that I've never forgotten. They didn't have to do it and they did it, and they never received any publicity for it. In fact, you're the first person that knows about it other than the immediate family.

### Oral History: Daniel Stefanides Jr.

Daniel Stefanides Jr. (figures 52 and 53) had reached his eighth year when his father became the youngest casualty of the Knox disaster. He recounts the personal and family's ordeal to recover from the loss. Clearly for his son, Daniel Sr.'s memory has survived the years. (From Daniel Stefanides Jr., taped interview, August 1, 1994, NPOLHP)

DANIEL STEFANIDES JR.: It was during the school year and I remember coming home from school not knowing anything had happened until I walked into the house. The first recollection that I had was seeing my mother and my grandmother, my father's mom, and several aunts sitting around with candles lit and praying the rosary. That's one recollection and after that it's kind of a blank. I have no recollection of who told me or exactly how it was told. My feelings were probably one of someone who was eight years old. It was tough to grasp the significance. I kind of remember [telling myself], "Well, he'll be home later." Obviously later never came.

I remember I didn't stay home that night. I stayed over at my friend Steve's house. Then it seemed that the next recollection I have is, we were up in Steve's room talking about it [and] apparently it must have sunk in. I just presumed that he was dead and he wasn't coming back. As kids talk, [we were saying] what we would like to do to these people who were responsible for this. I remember watching some of the news and I have a vivid recollection of their dumping the railroad cars into the river, attempting to stop the flow of water, and how the cars were just being tossed around and had no impact what-

Fig. 52. Daniel Stefanides Sr. holding Daniel Stefanides Jr., circa 1952. (Courtesy of Stephanie Stefanides)

soever on the hole. The river was controlling everything. There wasn't anything they could do.

ROBERT WOLENSKY: What did you want to do to the people who had caused the disaster?

STEFANIDES: Oh, I really don't remember thinking that there should be some punishment. In the context of being eight years old, what do you do? You throw rocks at them. When we were playing, we used to have a catapult where we used to shoot rocks and, you know, bang. Things like that.

As the days and the weeks progressed at first there were a lot of people around all the time and then gradually there wasn't. It wasn't a cutoff or anything. My birthday was two weeks later and I do remember people being over to the house for my birthday. I think that year my Uncle Richard gave me this huge model of a ship. That's strange that I can remember that.

WOLENSKY: Do you recall the memorial service?

STEFANIDES: Not really. I remember being over at the Court House. I guess there were some people that went to trial and I remember I was with my Uncle Steve. I remember him taking me over there because during the preliminary hearings mother testified.

WOLENSKY: What did this mean for the future? Do you recall anything related to it as you were growing up?

STEFANIDES: It softened, I guess, as with any transition. We did have a bit of an extended family. All my dad's family lived in Swoyersville a couple of streets over here and over there. I do remember spending a lot more time with them for a number of years. A single parent family was not the norm like it is today and I'm sure it was very difficult for my mother at times. As I got older [it got better]. But there was an extended family so I really didn't feel left out with the family and with the people in the neighborhood.

One of the things that I recall [is] that if we went down the other end of Swoyersville—which is maybe two miles—on your bike, people down there knew you as well as they knew you at the other end and so I never felt really isolated. You know, there may have been one or two times that, "Oh, jeez, I got to get out of here." But I really don't think that there was anything that I really missed.

There were times as far as direct parental supervision, you know

Fig. 53. Daniel Stefanides Jr., son of disaster victim Daniel Stefanides Sr.

being the oldest, I was supposed to assume a lot of things and many times I did not. But with Michael being a baby, my mother was more involved [with him] and he needed more care certainly. I often wondered what could have been if I had someone to play ball with all the time. You would like to think what your father would do for you, or if I would have been forced to study harder or anything like that. It may be my way of rationalizing it. Well, you certainly can't predict how things should have turned out.

WOLENSKY: Your Uncle Joe Stefanides worked at the mine. Did you ever talk to him about it?

STEFANIDES: Oh, no, no. It's one of those things. He lived on the next street over and he was my godfather too, but out of my dad's family we were probably least close to him. He has since passed away so you know there's no way of ever [knowing]. Certainly if you wanted a presumption, maybe there was some guilt and with guilt some people withdraw. I don't know. I remember hearing stories that it was on his encouragement that my father did go work in the mine. I'm sure you're well aware that they were scabbing [working non-union] and it's my understanding that the men knew that there was some risk involved, a greater risk than mining somewhere else. Apparently they were paid more money too. And my father wasn't making enough money at his business, a gas station.

WOLENSKY: Did the scabbing come up? Did your uncle talk about that or how do you know about that?

STEFANIDES: I think just from talking to some of the people in the area. I've had some contacts, not a lot—I haven't gone out and solicited [information]. But under the strangest of circumstances, I've met some people who knew my father. [For example,] I was out in Dallas at the prison and [met] the warden. I introduced myself the first time I was there and this gentleman said he was from Forty Fort and before the prison opened in the mid-sixties, my dad had a service station on the Avenue and he knew my father from that. We spent some time talking and I think it was him who first brought up the issue of scabbing. Some of the old timers in Swoyersville did too. I would presume that it was probably a non-union mine. The unions had been waning especially in the hard coal industry.

WOLENSKY: Mike Lucas was the miner in your dad's crew. Did you ever talk to him?

STEFANIDES: No, I've never talked to him. One of the things, and I don't recall whether it was true or not and I don't know the source, so it could be something that just evolved over the years, but someone had said once that my father was out and he went back down. Now, whether that's true or not, I don't know. He went out to pick something up. It could be something that has just evolved with time.

WOLENSKY: What has your mother said to you over the years when she has spoken about the disaster?

STEFANIDES: Very little, very little. I don't think I've ever really talked to her [about it]. Not that I can recall. My sister has a [video] tape of a Knox documentary.[1] They were home for a first communion and after the party Patsy brought it over and asked our mother if she wanted to see it. "No." She said, "I've never watched it. I have no desire to." [After watching the tape] I can only speak for myself but the more I heard the more I didn't know if I wanted to know.

Certainly there were things that were done wrong. I'm sure that's the case in any mining accident. But, you know, there were some blatant things that were done and the only thing that you can presume is that it was done for greed. I didn't even want to know. I know some people were fined and maybe a couple of them did some soft time. But it was suggested to me once that maybe it'd be better to pursue it and have some discussions especially with my mother. I've chosen not to. As a matter of fact, when I watched the tape I didn't handle it as well as I thought I would. It has piqued my interest, certainly, and maybe some day [I'll get more into it]. . .

I've heard a number of stories about my father, just some of the things that he used to do. He was a very well-liked man. I remember my Uncle Steve saying that if he had chosen to get into politics he probably would have done very well. My Uncle Steve was the district magistrate. For years he was the J.P. [justice of the peace] and things like that. One incident I remember, there was a field across from our house that Sordoni [Construction Co.] had and I guess over the years they had dumped rocks and broken concrete and things like that. He was the one that arranged all the kids in the neighborhood to go out and move all these stones and clear the field so we could have a place to play baseball. I guess he was pretty [well-liked]. He was an avid outdoorsman. He hunted and fished. I have some recollection of fishing with him and of it being very quiet and serene. I was having a feeling, a relaxed feeling, a secure feeling.

Michele [wife] was working at a little store out here and they were getting deliveries from UPS. The UPS driver started talking to her and she came home one day and she said, "I didn't know your father had a racecar." I said, "Well who told you that?" And after she mentioned it, "Yes, I do have some recollection of that." There used to be [race] tracks down at Bone [Stadium] and I even think they used to race down in Plymouth. This is before he went into the mines, when he had his own station. I think he learned to be a mechanic in the service. He was in the Navy.

WOLENSKY: Do you have any of his mementos?

STEFANIDES: Yes. I don't know where they are. I have some of his pins from the service, his hat, and let's see what else, Navy ring and that's about it, some pictures. One of the few things that I remember about the gas station, after a number of years, my mother had a stack of bills. She said, "These people didn't pay their bills, that's why your father went into the mines," and then she just threw them into the garbage. This was several years later. I do remember that. Apparently, he gave a lot of credit. It would probably go along with his good nature and trying to help people. "I'll pay you tomorrow," that type of thing. Uncle Richard said that dad stayed away from that [forcing people to pay up]. He did remember that. Richard has some pictures of all of them after the first day of buck [season], with cars covered with deer. Their old Buicks and those big cars. Some of your questions have pried some things open.

### ORAL HISTORY: AUDREY (BALOGA) CALVEY

The Baloga family lived close to the River Slope, a short distance from Mrs. Calvey's current home in Port Griffith. When the disaster struck, Audrey Baloga (figure 54) was fifteen years old and very close to her father. The grief felt by her and the Baloga family is apparent in these recollections. Like most other family members, Mrs. Calvey holds a strong resentment toward the owners. (From Audrey Calvey, taped interview, June 17, 1972, NPOLHP)

AUDREY (BALOGA) CALVEY: My dad was formerly from Czechoslovakia. He came over here, and my mom was born and raised here in Port Griffith. . . . When the water broke into the mines, I was just a young girl. I didn't feel well [that day] so therefore I didn't go to school. It was a cold blustery day. Sleet, snow, cold, windy. Somehow or other there was a lot of traffic going through Rose St. My mom looked out and she said, "Gee, there's an awful lot of traffic going by." I looked out and I didn't pay any attention to it and then I don't remember exactly how [but] we heard—I believe it was my uncle came over and said the river

Fig. 54. Audrey Baloga Calvey, daughter of disaster victim John Baloga.

broke in. And that's when we all got hysterical.

We found out that it wasn't far from where we lived. It was only about five hundred yards away. In fact, where we lived you could actually see the river. There was an early thaw up above in New York State somewhere and the river was on the rise. You could actually see the ice flows going down the river from where we lived, out our living room window. But we couldn't see the water going into the mines because it was on a curve. We went down toward that area and they said that the river broke in and there were men trapped. At the time they didn't know how many. It was in the morning around 11:00 a.m.

Then more people came and, of course, it was a coal mining town [so] just about everybody had somebody working in the mines there. My uncle lived next door to us and he came over and he said that they're going to get them out. Then my [other] uncle from up here on Main St., he came down and he said to my mother, "Caroline," he said, "We should go up to the hospital because they're getting some of the men out." Some men got out through the Eagle Shaft right down here, over the cliff here. I don't know the exact spot because I didn't travel down that area. Some men did get out and [went to Pittston Hospital], right up the hill from here.

Well, we went up there and they said, "Yes, he's here." They told my mom and they told me and we waited and waited and we went from room to room and he wasn't there. So then we happened to spot a missing miner's list and his name was second on the list. We came back home and we went down there again to the [break-in] spot. We were watching and waiting and waiting. It seemed like we were sort of not [exactly] losing hope but, I can't really explain it—it's that you knew something terrible had happened and you were hoping that it wasn't really true. We waited and then we had to finally go in the house because it was getting so blustery cold. Mom sat in the window and she waited for days and days.

Not one of the owners of that mine came by and told us they were sorry. Not one. Not ever to this day. My mom has passed away now but to this day not one of them or their families even sent a card or even acknowledged that those twelve men lost their lives. In my heart I say it's not the twelve men, I say it's another twelve apostles. That's the way I look at it. God had a reason for that.

There's just nothing we could do to change it but if men weren't so greedy with the almighty dollar this could have been prevented. Because the men were told to rob those pillars [of coal] and they robbed them so thin that the river had to break in sooner or later.

ROBERT WOLENSKY: Did your dad ever talk about what it was like to work at the Knox?

CALVEY: Yes, sometimes when he came home he would be all wet up to his waist and he said, "Well, we had to work in water today." His knees were all skinned up and he was frozen. He used to tell us—I'd hear him tell my mom—how crooked the owners were. They'd come in and they'd say, "Hey, you have to load so many cars." I was just reading one of my mom's letters to the editor where she said that he never made more than $19.95 a day loading thirty to forty cars of coal. So how could a person, a miner, get rich on that? And have a wife and children to support and bills to pay?

Every year across the street [at] St. Joseph's Church there is a Knox memorial mass, and there is a memorial monument of black marble. Every year we go [to the mass] and every month on the anniversary I light a candle. Because there still is no grave to visit. They never recovered their bodies. It's like a coincidence that years ago I bought this home and my husband said, "Maybe his body is under this house someplace in the mines." Who knows? But it would have been nice if we could go and visit a real grave, you know. I still can't rest over that. We never did recover any of the twelve Knox miners.

WOLENSKY: While your mom was at the window staring what were you doing?

CALVEY: Well, we cried too. I was the second oldest. My brother [Donald] is older than I am. We were just moping around. I had a younger brother to take care of because mom wasn't that well either, and we sort of banded together and tried to give mom support. But things never [were the same]. He never came back.

WOLENSKY: Did your mom ever come to terms with your father's death?

CALVEY: No. She died here in this home with me. She was sickly to begin with and she died July 1, 1980. I used to tend to her and we'd sit and talk and she'd say—"God how I wish your dad were alive to see this, to see that." And I'd tell her, "Mom, he can see us all from heaven." She [later] used to always say it was too bad that she couldn't go by the cliff over here, but it was too dangerous to take her to see anything. She never did get over it. Right after the Knox disaster, when she finally accepted that he was dead, her hair turned almost completely white over night. It was white, snow white. The doctor did say that was the cause—dad's death. . . .

My heart was broken when I finally accepted that he was dead. You know you hear about miners getting trapped in other areas and you never think that it's going to happen here. Not to your own dad, no, not my dad. But it did and my heart was broken because I was very close to my dad. I was very, very close. I would wait for him when he'd come from work and I'd have a bottle of beer for him in the summer time and while he was drinking the beer, he'd say, "Hey Aud, take off daddy's boots," and I'd wiggle them off his feet for him. Of course, they were all wet and muddy but I didn't care. And he'd say, "Well, could you help me out of my jacket?" and I'd help him out of his jacket. Then I'd help get his water ready for his bath. I have very fond memories of my dad. I was really very close to my dad. If he was going someplace he'd say, "Come on Audrey, let's go," and we would go together.

I was about fifteen or sixteen years old back then. I wasn't that young that I didn't understand anything, but I was a young girl at the time. I had to quit school and go to work to help support the family. My mother had relatives and they got me in at MK [garment] Manufacturing Co., and they taught me how to sew. I was put on floor work and I've been a seamstress ever since. 'Til this day I still work as a seamstress. And I love it. . . .

WOLENSKY: You did become angry about the whole thing?
CALVEY: Yes I did. I thought that if I ever met the owners of this mine I would want to kill them myself, I really would. To this day I think I would slap their faces if I ever saw them, that's the anger that's in my heart over this. The least they could have done was come by [to] offer condolences or even offer a mass. Nothing. Not a thing. But they could still live in their high mansions. Most of them are dead now, but they did all have beautiful big homes. . . . I have learned to block [a lot] out of my mind. It's still there but it's just like in a filing cabinet put away, locked away so that there are some memories that I just don't like to pull out.

## Oral History: Donald Baloga

Donald Baloga (figure 55) sped to the River Slope Mine in hope that his father would be among the rescued. The oldest of four children, he assumed the role of provider upon his father's demise. Ironically, as a young boy he and his friends swam in the river at the exact spot of the Knox breach. As with some other members of the victims' families, he reports two inexplicable occurrences associated with his father's passing. (From Donald Baloga, taped interview, June 20, 1992, NPOLHP)

DONALD BALOGA: I was born in Port Griffith in 1938 and raised all my life in Port Griffith. We moved around a little until we settled on Rose St. The Knox Coal Company mine was located [there]. My father worked for the Ewen Colliery [of the Pennsylvania Coal Company] and they went out of business so he took a job with the Knox. Of course, my father wasn't educated that much. He told me he started when he was thirteen years old in the coal mine. . . [and] worked maybe twenty-six or twenty-seven years. He used to walk from Port Griffith all the way to Scranton before I was born to go to work, and walk back. Up seventeen miles and back. They had no cars. He'd start out two or three o'clock in the morning to go to work at 7:00 a.m.

He was a good father. He taught me a lot of things. He was a strong man. He wasn't built big but every inch on his body was muscle. . . . He was a hard worker, never missed a day of work unless he was really [so] sick that he couldn't walk. He had arthritis, rheumatism. He even applied for a pension for black lung, miner's asthma they called it at the time, but they told him that he would probably only collect $37.00 a week. He said, "I can't live on $37.00 a week. I got six people to feed." So he said, "I'll work 'til I go down."

He took the [Knox] job. It was a rough job and he always told me that he feared the river because he said that the Knox bosses knew they were mining coal under the river beyond the red line. And the inspectors, nobody knew what happened to the inspectors, if they were inspecting or not. My father always told me he said, "Someday that river is gonna break in and if it breaks in we'll drown like rats. . . ." He kind of had a premonition. . . .

ROBERT WOLENSKY: Where were you when you first heard about the Knox break-in?

BALOGA: Well, I was in our garage changing a headlight on my car, I'll never forget that. My mother screamed out. I heard her screaming and I went to the window. She said, "The river broke into the mines [and] your father's down there." I said, "Oh boy." I hurried up and jumped in my car and I went up to the [May] Shaft on Friend Street. There was a breaker there and that's where people were standing around. They said the water was up to the top. I knew there was no hope. I heard somebody say there was a man by the name of Fabrizio—he was the [one of the] owner[s] of the coal mine—[who] came up and he said, "Oh my God," he said, "my mines are ruined." And everybody said, "How about our fathers down there?" That's the kind of man he was. I was really mad. I didn't [actually] hear it. People told me that he said it. I tried to see where he was because at the time I think I would have choked him. He was on vacation and he came back when they notified him that there was a disaster in his mines.

Fig. 55. Donald Baloga, son of disaster victim John Baloga.

I hated to go back and tell my mother that the water was up to the top, [that] there was no hope. Then I heard somebody saying that there was a shaft down by the river. They might get out that way. I knew where the shaft was. It was an air shaft right below the hospital by the Lehigh Valley railroad tracks. So I went down there. Nobody came out and then a couple of [hours] later the guys were coming out. I said to one man, "Is my father [okay]?" "Oh yea, he's right behind me." The last man came up and I didn't see my father. So somebody was lying to me and I felt real bad about it. If I seen those mine owners I probably would have shot them all, that's how bad I felt. . . .

I went up to the hospital looking for him but he wasn't there. I looked at just about every man there and I asked if they saw him and they said, "Oh, he was right behind us but we never saw him [again]. He never came out. . . . "

Well, I went home crying. I was telling my mother and [she] was really tore up. She didn't sleep right. She was sick from the day he died. [From] the day he got killed in the mines she was always sickly. She never recovered from that fully, I don't think. She must have been about forty-two then. Every day we used to walk and try and get some news [if] anybody came out of the mines somewhere. No, nobody. . . .

One thing that really gets me is where it caved was where we dove and swam. The same hole. We used to see air bubbles come up when we were swimming. I learned to swim in the river and we used to swim there often. That spot where that hole went in was the exact spot where we had cleaned out the rocks so we could dive deeper.
WOLENSKY: How many rocks did you remove?
BALOGA: Oh, just the loose rocks that we could get out of there. We didn't dig in but any rocks that were loose we got. A lot of times we pulled out rocks maybe half a foot [in diameter]. We used to see air bubbles come up from the mines. "That air is from the mines," I said to the guys. I said, "Yep, I'll bet any money there's some holes in that mine somewhere, air coming up from the mine." The exact spot. Right to the tee it caved in the same [place]. . . .

We had a double deck diving board. It was built out of old railroad ties from the Lehigh Valley railroad. They'd throw the old ones on the side and we used to float them down all the way from the Falls Bridge and then we'd pile them [and] put them on the shore. We had the only double deck diving board on the river. I'd say it was twelve feet high from the top of the board down to the water. We had two boards, one on the bottom one on the top. Charles Staliha built it for us. We [all] helped build it. We used spikes and sledgehammers. It survived one flood [but] then the next flood [1972] took it. We used to have rocks piled around it and everything. The very spot that accident happened. . . .

Could you imagine if that happened in the summertime when we were swimming? There was about fifteen or twenty of us swimming in there. We used to swim bare ass—no trunks on. That would've been fifteen kids gone, sucked right in. I never saw a suck hole that big when that happened. I watched the [railroad] cars go in and heard them mine cars crunching.

I think they said three million gallons a minute was going in that coal mine. I don't think my dad had much of a chance. Some of the men said that they saw him get it fast. He got washed in. Some said they told him the river broke in and he went back for his tools. But I doubt my father would go back for his tools because he often told me, "[If] that river breaks in we'll drown like rats," and he'd run. . . . Maybe when they hollered the river was coming in he ran toward the shaft. One guy said he ran to the shaft—he saw him—and then he ran back for his tools. I said my father would never run back for his tools, not when he knew the river was coming because he always sensed that. He had a premonition [for] a long time that the river would come.

WOLENSKY: Did he talk about that often, Don?

BALOGA: Yes he did. I hadn't heard him mention the river for quite awhile, maybe a year or two, and then the day before that happened he said to my mother, "Boy, that river is pretty high. Them icebergs are coming down." He said, "I fear that river." He had that premonition. I heard him tell my mother. They were talking. . . .

WOLENSKY: Do you think they could have prevented it? Do you think they knew about it?

BALOGA: Oh, they knew about it. Sure, my father said that he told one of the mine foremen that we were going beyond the red line and the mine foreman said, "You got a wife and kids. You keep your mouth shut. There's guys looking for a job." That's what my father told my mother and I heard it. I was at home at the time. . . . I guess a lot of people weren't surprised in Port Griffith when it happened. I'm sure that most of the miners [knew] from what I've heard at different times. They knew. . . .

WOLENSKY: Did your father ever work in any of those illegal chambers beyond the red line?

BALOGA: Oh yes. He had to. Many times they mined beyond the red line. Not only my father. There was other men went beyond that red line. . . .

WOLENSKY: You have to ask yourself, how could experienced miners, men who had mined all their lives, get that close to the river?

BALOGA: They had no choice. Either that or lose your job. I know that was it. Guys had to work for a living, they had big families. Some of them didn't know anything else. My dad had no experience anywhere else. He had to work in the coal mine. That's all he did all his life. Most of the guys in Port Griffith, that's all they did all their lives. And if they complained to the bosses, my father said that the bosses would say, "Hey, there's a lot of guys out there looking for jobs." My dad was laid off for quite awhile. It was hard collecting unemployment checks but most of the men that worked there had no choice. I'm sure that he wasn't the only man that complained about the rock cover. . . .

WOLENSKY: Did you ever hear from any of the owners?

BALOGA: No, I never did. . . . Never seen them, never talked to them. Didn't want to because I think if I did I'd probably want to kill the guy. That's how bad I felt about it. But they stood clear of the family. They never came down and apologized or anything. Never got anything from any of the owners. They never sent anything to families, not even a card. They were just a bunch of men who were hungry for coal and it didn't matter if the guys got killed or died. They wanted that coal out of there. It was bound to happen sooner or later. . . .

We went to court and my mother sued in Lackawanna County and the others sued in Luzerne County. There was eleven families [that] sued. One of them might have been bought out, never sued. My mother only got like $13,000 out of the deal. They couldn't prove suffering or anything, you know.

WOLENSKY: Why did you go to Lackawanna County?

BALOGA: Well, I felt it was a better court than in Luzerne County and she did get more money from Lackawanna County than Luzerne County. And having it in Luzerne County was too much publicity and all that. Our lawyer in Scranton was Christopher Powell. I think he's deceased now. He said [it's] the best thing you can do. You couldn't sue for suffering or anything like that. We got $13,000. That's a drop in the bucket. No money in the world could replace a life. But you got to take what you can get because my mother had the four of us to raise.

WOLENSKY: How old were the kids at this point?

BALOGA: I was twenty-one, my brother was about eight or nine, Audrey was about sixteen, and my sister Sandy was about thirteen. I was the oldest at twenty-one. My brother hardly remembers my father at all. . . .

WOLENSKY: Did you have to go to work?

BALOGA: I really had to go to work because I wanted [to help] my mother. I didn't want the money to disappear. I had been working at Duchess Apparel at the time [of the disaster] but I was laid off. They found out about the mine disaster and they called me back. They helped me out and I worked there for five more years. Then I went to Marvel Industries driving a truck and worked for Marvel about six or seven years. . . .

Everybody knows it was human greed in Port Griffith. Everybody. There isn't a miner in Port Griffith that worked for Knox Coal Co. that would tell you different. I was close to my dad. I always felt bad about it. I never went near the river again. And you know it took my dad and then in 1972 [Agnes flood] it took my house in Plains, the river [did]. Yep, took my house. We got a grant and we bought our home in Dupont. I think it was time enough for me to move away from that river, and I loved the river. I used to fish there, swim there. Oh boy, we used to have a lot of fun in that river especially on that diving board. . . .

WOLENSKY: Your mother never remarried, did she?

BALOGA: No, didn't want to. She said there was only one man for her—him. That was it. She was sick for a while then she got better but she never recovered after that completely. I used to keep her happy. I'd make her laugh and all. The doctor told me [that if] it wasn't for me and my sense of humor she wouldn't have made it two years after that. I used to keep her laughing all the time. We'd take her all over and I had a hell of a sense of humor at the time. I'd make everybody laugh, even make priests laugh at me. But then after I got married she started getting sick. She was a heavy woman, too heavy for her feet. High blood pressure, gall bladder trouble, and then she got cancer. Doctors played around with her. They told her that she had woman trouble and said it was a cyst growing. They hit it with a laser or something. She came back and the doctor said they got it all, it's okay. Then she started getting sick again, same thing. Went back to the doctor and he said he'd send her to the Philadelphia hospital. That doctor told us there's nothing you can do—the cancer's all over her. She died in June 1980. . . .

The day before the mine disaster happened—not many people know about this—but my father was the type of man [who] never liked to be late. He set his alarm clock [but] the alarm clock didn't want to run that night. He shook it [and] went back to sleep [but] he couldn't sleep because he was worried about being late. He told my mother. I heard him tell her that there's something wrong with that clock; it won't run. Shook it, it'd run for a while and then it'd quit again. Never had problems with it before. He didn't get much sleep, I know, because I could hear him off and on walking around. I guess he was worried about going to work too late. So he went to work and the disaster happened.

Then after that disaster happened, my father had a seven-day clock. He always wound it and after that disaster happened, it was eleven days and I told my mother that the clock is still running. She said, "I know he always used to get mad if he'd forget to wind it." It would run out after seven days. And that clock ran twenty-one days! Then it stopped. I watched that clock and I said to my mom, "I ain't gonna touch it because maybe that's how long he has been alive down in the mines." He might have got to a high spot because my dad knew every inch of the mines. He told me that he could go into the mines in Pittston and come out in Scranton somewhere. . . .

Strange things happened. We talked to the priest about the alarm

clock and the big clock and he said, "Well, the big clock might have been the days he was alive, twenty-one days." After twenty-one days it just quit. It was a seven-day clock because it said it was a seven-day clock on there. He had it fixed a couple times. [It was] a big old grandfather clock. That was his pride and joy, that clock. He'd wind that thing all the time, shine it up. Boy, that looked nice. And it ran for twenty-one days. The priest said that [the alarm] clock that didn't run was probably a warning to try and get him not to go to work.
WOLENSKY: Was that alarm clock broken for good?
BALOGA: That clock ran after that, yes it did, because I used it to go to work.
WOLENSKY: Did you have it fixed?
BALOGA: No, there was nothing wrong with it. I'd wind the clock up after that disaster and it'd always run. It never quit. Wouldn't work that night, the night before, no. It wouldn't work.

### Oral History: The Ostrowski Children

When their father, Eugene Ostrowski Sr., perished, Eugene Jr. was twelve years old, Donna was five; and Anita Ostrowski Ogin was fifteen months (figures 56, 57, 58). They recall their father's death as well as their mother's difficult but successful efforts to provide a normal life for the children. (From their taped interview, July 29, 1992, NPOLHP)

EUGENE OSTROWSKI JR.: I knew my dad fairly well. Being I was the oldest we used to pick coal together, fish, hunt. I only hunted one year legally with him because I was twelve in September so I hunted with him in the 1958 season and that was the only year we hunted legally together. . . .

He worked [mined] in Wanamie. He also worked in Avondale, which is in Plymouth. He worked under the river there also, in three feet of water. . .but they were going straight ahead and they'd keep pumping it out. He started off in Glen Lyon. . . . He went to Jersey, worked in the Bake-Rite plant for a year and a half. But he didn't care for the city and he came back and decided to go back in the mines. And then in the mines, they weren't getting the work so he decided to go down to my uncle, Stuart Creasing, who was a rock contractor. He had gotten him a job in soft coal mines down on the outskirts of the Monongahela River by Pittsburgh. Then he picked up the job here at Port Griffith. . . .

Now I slept with my dad. He and I slept in the back bedroom [in] which there was no heat and I mean beer bottles would actually freeze and crack—that's how cold it would be out in the hallway. We had a big thick—we called them *pierzyna*—a [Polish] feather quilt—and [we'd] crawl underneath that and breathe heavy for about ten minutes and then you finally fall asleep. But you can't stick your nose outside or else you get frostbite!

He'd wake me up in the middle of the night and he'd be showing me the cracks in the ceiling because we had old, plastered ceilings and they were all cracked up. He'd tell me, "Jeez, the ceiling's cracking, the roof is going to fall, the roof is going to fall," and I'm staring there and I know those cracks were there before. But he had dreams of the mines because he'd see cracks in the ceiling of the mine and that would bother him. . . .

ANITA OSTROWSKI OGIN: Wasn't it like three days before the cave-in, that daddy was telling mommy something about, "You can hear the water running above us?"

E. OSTROWSKI: Yeah.

OGIN: [He said], "As we're working you can hear the water running so we know we're close to the river." That was the only thing that I remember my mom ever really saying. He was telling her that he can hear that water. . . .

ROBERT WOLENSKY: How did your mother learn the news?

E. OSTROWSKI: That I'm not sure because I was in school. I was in seventh grade at the time in Pulaski Junior High School in Glen Lyon and they just pulled me out of class. They said, "Step outside in the hall for a minute and you have to go home." I went outside and my uncle was there to pick me up. I'm pretty sure it was my Uncle Stan and he never said anything [about the disaster], he just said, "Well, you have to go home." I didn't find out what was what until I actually hit the house and my mom was crying and everybody was worked up. Maybe one or two of the neighbors were over at the time but that was the first time I found out about it. When my mom found out, I have no idea. . . .

I know he did not come home that night. He was supposed to be

Fig. 56. The Ostrowski family, 1958. From left to right: Donna, Mrs. Theodosia, Eugene Jr., Eugene Sr., and Anita. (Courtesy of Ostrowski family)

home roughly eleven or twelve o'clock but she probably figured that he had worked late, maybe an extra half a shift or whatever. By the next day already it was in the news that the river had flooded the mines. That's when she knew where he was. She was not sure that he was in that section of the mine but nobody knew anything. It wasn't until later that we found out that he was one of the ones working where the river broke in. All the other guys that escaped were in dif-

Fig. 57. The Ostrowski family, photograph taken immediately after the disaster. From left to right Eugene (age twelve), Mrs. Theodosia Ostrowski, Anita (age fifteen months), and Donna (age five). (Courtesy of Ostrowski Family)

ferent chambers and they had a chance to find a way out.

WOLENSKY: Did you head up to the site?

OGIN: From what I understand, our uncle, my mother's brother, wouldn't let her go. He said, "You stay here with the kids," and he went up. He was staying up at the mine and then every now and then he would come home and give updates. I remember my mother always saying that our uncle would not let her anywhere near it. He said, "That's no place for you. You stay with the kids, I'll go." I think other relatives came in, *Ciocia* [Polish for aunt] Helen and everybody. Everybody started coming.

WOLENSKY: Gene, you didn't go up at all?

E. OSTROWSKI: No, we went to pick up the car, which he used to leave parked at where the Hanover Nursery is now on San Souci Highway. My uncle and I and my mom went up to get the car, because he'd park it there and he'd ride with a couple guys. I'm pretty sure he rode with Tiny [Gizenski] from Muhlenburg. . . .

WOLENSKY: Can somebody tell us about your mother?

OGIN: She always said that it was her strength that got her through everything, her strength and the three of us. But personally I don't know how she ever really did it. She had to have had some kind of a very strong inner faith and be a very strong person to be able to do what she did, to raise the three of us and everything. We were a very close family.

DONNA OSTROWSKI: Yeah, I think about that now.

E. OSTROWSKI: I'll tell you one thing, mom would give up anything. She would rather starve first [than have us go without]. If she was going to take us out for ice cream to the Carvel stand, and if the neighbor's kids were over or whatever, everybody goes. She wouldn't slight anybody. "We're all going." She would sooner do without herself than to slight somebody else whether it was one of our friends or whatever. . . . She was born in the house in Wanamie. Her maiden name was Gorczyca. Theodosia Gorczyca. Tozia is what people called her. . . .

I remember at first when it happened, there was a small church, St. Mary's Lithuanian Church. She was over there almost constantly because that was when *Ciocia* Lillian, our Aunt Lillian, came in and she stayed with us to baby-sit while my mother was in church all day. And then maybe she'd come home and have something to eat but then she'd go right back again, until they would lock the doors. She kept feeling that [because] she was praying maybe something good would come out of it. She kept having this idea, too, that maybe he crawled up one of the air shafts and he was up there, that he was safe and he was waiting for the water to go down. . . .

She figured that they'd find two or three stragglers. They'd come out different air pockets from different mines instead of being in Port Griffith. They were clear out to maybe Plains [because] they worked their way through these old tunnels, old workings, and they'd come out far away from the mine. She had a feeling that maybe there was a possibility that he could have found a place to get out. But then after a week or ten days that was about it. After that it was just like, it'd have to be a miracle. So automatically you just figure, well, you start to face reality—he's not coming [home], he's not in any air pocket and if he's trapped there, he's dead. You know it's just—he's gone. That's it. Face it.

OGIN: What she also did was she kept a scrapbook and she kept figuring that one day if he did get out maybe he would have amnesia or something and some day everything would come back to him and he'd come back home. She was going to pull this scrapbook out and say, "Look at all the ruckus you caused!" So she kept all the clippings out of the newspaper and that's what I have here. . . .

WOLENSKY: Were times tough after that?

E. OSTROWSKI: Oh, very tough, very tough, very tough. Clipping coupons, it wasn't a habit or something, she had to do it. I picked coal. [Earlier] I picked coal with my dad because we had to; we never had a furnace. We had two stoves, one in the kitchen, one on the far left. I picked coal since I was probably eight or nine years old and even after he died I still picked coal until I was about eighteen years old and then she finally got a furnace. That was our source of heat—she wouldn't have to buy a ton of coal. When he died people were nice. They sent food baskets and a free ton of coal, which I liked to keep on hand. We put that on the side and wouldn't touch it [so] if I had a basketball game or football game to go to I wouldn't have to pick coal or crack coal in the middle of winter. That was a spare.

WOLENSKY: Did you ever hear from the mine owners?

E. OSTROWSKI: No, no, no. There were officials from the mine union, United Mine Workers, that came over who brought boxes of food, canned goods, powdered milk. The Economy Store did give us a canned ham. There were donations for that. The United Mine Workers were the only people that we ever heard from. As far as anybody with the [Knox] mine, no. Anybody that owned the mine or part owner or whatever, nobody ever [contacted us]. There was nothing.

WOLENSKY: I'd like to ask all three about your feelings toward the mine owners. What if you could talk to them?

E. OSTROWSKI: Well, I'll tell you what, I feel this way. My dad was involved with roadwork. He didn't want to leave town. He could've had a good job operating equipment [but] he didn't take it. He had a job in Jersey in a plant [but] he didn't like the city [so] he came back. No matter what they did, I really don't have hard feelings toward them. Okay, so they were trying to rob coal, they're cutting up the wrong angle, and they hit the river. But I myself, I don't blame them. I blame him. He wanted to work in the mines.

WOLENSKY: Did he like mining, Gene?

E. OSTROWSKI: Not necessarily, I don't think. At that time that's where the biggest money was without leaving the area. He'd rather stay in the area so I really don't blame them. I probably blame him more than I do them. That's my feeling.

WOLENSKY: Donna, what's your opinion?

D. OSTROWSKI: A lot of hatred for the mine owners. There were so many times especially around that time of the year, around January, that I would start feeling so depressed and break into tears. I never wished anything bad on anybody, but for those people I just wished that they would die like river rats. That's all my prayers would be. And when I saw in the paper about one of them, I don't remember which one it was, it was just recently like within the last two years, and it had about how he was suffering. He went from one hospital to another. . . . But to me I was almost like, "Thank you, God, thank you," and I hoped that he just continued suffering. They were trying to do everything for him to help him and it was only because he was such a figurehead with the Knox mine. I feel that these people didn't care then for these families with small children, they didn't offer any type of assistance or even a little "I'm sorry," something like that, or "Here's a bouquet of flowers." Nothing at all. That's why now when they're in their elderly years and may be dying from cancer I couldn't care less about them. I just hope that they suffer.

WOLENSKY: Anita, what do you feel about it?

OGIN: You'd have to hit that pause button [on the tape recorder] too many times! I don't think they should call it the Knox Mine Disaster. I think they should call it the Knox Mine Murders because I feel as though everybody was paid off, everything was under the table and hush-hush, and double dealings. I just feel like it was an out and out murder type of thing because they knew they went past a certain point that they weren't supposed to. They changed maps later and so . . .

D. OSTROWSKI: Nobody wanted to claim ownership to it. Now what was it—these people owned the mine, they had the workings right there, but yet after this happened, "I don't own it!"

OGIN: Uh huh. . . .

D. OSTROWSKI: They were just greedy that's all.

OGIN: Greedy.

E. OSTROWSKI: But they had been doing this all those years. Every coal contractor around did the same thing. I don't care if it was Corgan, Biscontini, I don't care who it was. They were all crooked. . . .

WOLENSKY: Was your mother bitter?

E. OSTROWSKI: She was bitter all the way around. I mean who wants to lose their husband. . . ?

WOLENSKY: Would she go to the Knox memorial masses at St. Joseph's Church?

D. OSTROWSKI: Oh yes. We'd all stand there. They're reading the names and we'd be alright until they would call daddy's name and then you could hear somebody—sniff, sniff—and then before you know it was just like a domino effect right down the whole pew. I know when we first used to go up there to the mass I think pretty much through the whole mass my mother would be quivering and yet she tried to block it out and [say], "I'm okay, I'm okay." She never really liked to cry in front of us.

[Sometimes] she would say, "Oh, I'm just feeling blue," and she'd go into the bathroom and you'd hear her crying in there, but she never wanted to cry in front of us, it seemed. In front of us she was always the tower of strength. By the time I got older I remember

Fig. 58. Anita (Ostrowski) Ogin, Eugene Ostrowski Jr., Donna Ostrowski, children of disaster victim Eugene Ostrowski Sr.

asking her one time something about, "What happened with daddy?" and I hit her on a day when maybe she was in a blue mood. Well, when she started crying I figured that was the last time I was ever going to make her cry. When I did get older by that point it was kind of like water under the bridge.

WOLENSKY: What would she tell you when you were five and six and seven about what happened to your dad?

D. OSTROWSKI: Just that the mine caved in and the mines flooded and that he went into work and never came home.

WOLENSKY: So she would tell you straight?

D. OSTROWSKI: Straight out. I was five [when he died] and I remember it was in the afternoon, it was just starting to get dark then . . . I remember everybody was crying but she was pacing back and forth. It was getting later and everybody was upset and that's when the six o'clock news came on and I'll never forget that because we had the box of tissues right there. She wouldn't let anybody say anything and that's when they came on and talked about [the disaster]. But before that, every time she would call up there they kept telling her, "Yeah there was a slight accident," but they wouldn't give her any details. They wouldn't give any names or anything. . . .

WOLENSKY: Did you ever have a funeral or any kind of service to commemorate his death?

OGIN: Just the mass at Port Griffith every year.

D. OSTROWSKI: And we would get the masses at St. Michael's Church. She would have a mass offered right around the anniversary.

OGIN: I think it was about ten or twelve years later that she found out that there could have been a service where there would've been a coffin and a burial. That way it would have kind of like [ended] any hope in her mind. In other words she would have let him go. But by that point, why go through it now he was declared legally dead?

E. OSTROWSKI: That's another thing—at that time you were not legally declared dead until after seven years. . . .

WOLENSKY: Were there any unusual occurrences associated with his death?

OGIN: Well, when my parents were first married they lived in a poor apartment home, in the upstairs apartment. My aunt and my grandmother lived in the downstairs. Over the years from my aunt living downstairs she knew my father's walk when he would come in from work or whatever. She knew what his footsteps sounded like on the ceiling. So two days after the mine caved in my aunt was staying with my mother and trying to help her out and she was sleeping on the couch in the living room. She said that she heard my father's footsteps. She opened her eyes and she could hear those footsteps going right past the couch right into the bedroom. She opened her eyes and made sure she wasn't dreaming or anything. She's looking and you could see from the street light that nobody was in the room yet she heard these footsteps going past. She heard something in the bedroom and she went in because she thought my mother was crying or whatever. She went in there and she said everybody was asleep, everything was quiet. She always said that my father came back for a last goodbye. This is my Aunt Lillian Holton. . . .

D. OSTROWSKI: That I remember. And then there was one dream that I had. I was, let me see, about seventeen or eighteen and I was sleeping on the couch in the living room. I don't know why I wasn't in the bedroom but I had the night light on and I also always kept the flashlight right next to me wherever I slept. I heard his voice like whispering my name, "Donna, Donna." At first I thought it was my mother so I picked my head up and when I did I still heard his voice. My mother had this big wicker rocker in the living room and it was about four feet away from the couch. I noticed with the nightlight being in the wall right behind the rocker that the rocker was moving [to and fro]. It's not like the windows were open so there was a draft, and it was far enough away from me that I couldn't have bumped it in anyway. It didn't frighten me because I knew it was him. It was his voice. But that's when I started thinking maybe my mother was calling me. So I got up and I ran in the bedroom and looked and here she was sleeping.

I'll never forget that. The next morning it upset me so badly that I just kept crying. I felt as if I couldn't talk to him, you know, that he was there and he just came and left and that was it. I kept telling mommy about that dream and she was trying to calm me down because she knew how much it upset me, but at the same time I wasn't afraid or frightened.

WOLENSKY: Have you ever gone back to the site, right up to the river's edge?

OGIN: Yeah. I'm not even sure if we really found the exact spot. I was going from memories of what I saw in the news clippings and the way the river was shaped. We parked right when you get into Port Griffith in the alleyway, my husband and my daughter and myself. We just went for a ride and I said that I wanted to see where he walked into work that morning. Well, we had no idea. We just started walking the railroad tracks and as we were going, it was really a stinking hot humid day. At one point all of a sudden you could feel this cool air real quick as you passed. We backed up and it was just like a rock ledge there. I couldn't see any openings yet we thought well maybe we passed an air shaft, maybe there was something around there. After a while there were these steel pipes sticking up along side the tracks and we just assumed that they must have been air pipes. They were all filled in and we kept walking until we saw what it looked like on TV, the way the river was shaped and how they were just shoving the railroad cars in. So we were in that general area. I always wanted to contact somebody that actually lived in the area and ask them where it was. I know they've closed it up since then, but I still wanted to see exactly where he walked into work that day. . . .

WOLENSKY: Did you go back Donna?

D. OSTROWSKI: Yes I did. It was behind the old Pittston Hospital up and over the hill. The day when I was up there I was looking and I started walking the railroad bed. This older guy came up and over the hill. I started getting a little bit afraid because I didn't know who he was, but that's when he asked if I was looking for something. I said, "Well yeah, I'm looking for where that Knox Mine Disaster happened." He said, "Ahh. Why would you have such interest in it? It looks like you're not old enough to really remember it." I said, "Well, my dad was one of them." And that's when he seemed to soften up a little and he said, "Right there is where the hole was, where the water went in." But you know as far as anything being there now . . . you wouldn't know that anything was underneath.

E. OSTROWSKI: No, you couldn't tell by looking at the river.

D. OSTROWSKI: But he did point out the spot.[2]

## Letter to the Editor: A Victim's Son Wants Answers

(From "Son of Missing Miner Wants Answers from Knox Owners," *Wilkes-Barre Record,* January 28, 1959)

I have paced the floor to find words to start this letter, but it was very hard knowing my dad is trapped in the Knox Coal Company mines. We still have faith and hope in God that they will come out alive. I will not rest until I know all the answers to a few questions I have to ask.

1. Is it a mining law that mines should have escape outlets? If so, where was the escape outlet at Knox Coal Company? If it was Eagle Shaft, why was it covered with debris?

2. Did the mine have any warning signal?

3. How come the tunnel was only 30 feet under the Susquehanna River? Was this the safety limit? What amount of rock strata is under the river and how far down do you have to go to be sure that it is safe for mining?

4. What is the exact complete ownership-managerial setup of the Knox Coal Company? Is someone hiding something? The owner or owners should have had the common decency to visit each family of the 12 trapped men to assure them that there is hope for their bread earners. Don't bother now because you have showed your true side.

5. This is the last question I have at the moment. How come the long delay in pumping operations? This is Monday, January 26, 1959. The disaster occurred Thursday, January 22, 1959? Where are the pumps? We fully realize the hole had to be blocked, but couldn't pumping operations start immediately at the shaft as the water was rushing in? That would mean much less water to be pumped out now. Every second and inch is precious for those trapped men. From newspaper reports the pumps at hand pump 5,000 gallons of water per minute. That would be 300,000 gallons per hour and 8,200,000 gallons every 24 hours for one pump. Multiply this by the available pumps in the Valley.

These questions must be answered with absolute truth.

The only way to avert any more mine disasters of this magnitude is for the workingmen of Wyoming Valley to ban together and demand the truth. Be men and don't be afraid to speak the truth.

Maybe what is needed is for the women to take over as Min L. Matheson has done for the dress industry. Maybe she is what we need.

God be with the trapped men and their families.

We deeply appreciate the help and kindness showed by all the Wyoming Valley people. If names had to be printed the newspaper would not have enough room for them.

—Name Withheld by Request

## Notes for Chapter Five

1. WVIA-TV, public television in Northeastern Pennsylvania, first aired the Knox Mine Disaster documentary in 1989. Co-producers: Ray Pernot and Ed Finn; director: Ed Finn; narrator: Erika Funke.

2. Members of the Greater Pittston Historical Society (GPHS), under the leadership of Bill Best, John Dziak, and Bill Hastie, examined old mine maps with survivor Joe Stella and found the exact spot of the break-in. In May 2003, the GPHS placed a granite marker at the site which reads, "The Knox Mine Disaster Occurred On This Spot, January 22, 1959." Fr. Hugh McGroarty blessed the marker in a dedication ceremony. As part of the forty-fifth anniversary commemoration on January 22, 2004, several persons braved the frigid temperatures and snow to visit the marker and pay their respects. See figure 72 in chapter seven.

## Chapter Six
### Remembering the Knox Mine Disaster: Poets, Composers, and Writers

*The Susquehanna gushed with a monstrous flow into the mine. Frank would never be found.*
    Erik McKeever, "The Day the River Came in the Mine"

*To the men who work below, Port Griffith, Pennsylvania, where the Susquehanna flows.*
    Frank and Tom Murman, "Port Griffith, Pennsylvania"

Northeastern Pennsylvania has enjoyed a long tradition of poets, composers, and writers who have celebrated and memorialized various aspects of the region's history and culture. The Knox Mine Disaster has become part of the tradition. The following items, composed between 1959 and 2004, capture the strongly held feelings toward the catastrophe. They have been arranged in chronological order to illustrate the enduring effect of the Knox disaster on the collective memory of community.[1]

## Poem: "Twelve Men Die"
### by Rose Chickson of Wilkes-Barre

If my Dad were working there
And heard a roar, he'd stop to stare,
And seeing tons of water gush
In a powerful and foamy rush,
He'd scamper to much higher ground
Not where the water's path was bound.

But if he helplessly stood below,
And caught the water's first mad flow,
He'd have no time to turn and flee,
He'd stay afloat, though hopelessly.
But say he ran to safer ground,
And prayed for days that he'd be found.

What torture must a trap [sic] man bear,
When he's beyond all hope and care?
My Dad was good, a quiet man,
Who sacrificed his full life span
To earn our daily bread and keep,
Now he's entombed—eternal sleep.

Don't pity him, his troubles end,
Beyond the reach of foe and friend.
And in my heart he'll always be,
The perfect Dad he was to me.
Oh, men, who betray life and trust,
You, too, will some day turn to dust.

Fig. 59. "Mine Flood is Halted," Wilkes Barre *Sunday Independent*, January 25, 1959.

### Poem: "Disaster"
#### by P. J. Nelson

The waters came in silent, murky flow,
And spread their liquid fingers far below.
A torrent uncontrolled; a rage unbound,
That paved a path of death beneath the ground.

Nature's malice; nature's heavy hand
Had struck beneath, instead of overland,
And spent her fury deep within the caves
Where men sought coal, to find but sodden graves.

A flash of chaos, swirling silvered fate,
While hearts above, with fearful hope must wait
And waiting done, could only turn away,
To seek the dreams that vanished here today.

Beyond the world of mines and mortal men
Above the reach of tearful human ken,
Some great Intelligence says, "Trust Me,
I have my reasons why such things must be!"

### Poem: "Mine Disaster"
#### by Mrs. Reese of Plymouth

(*Times-Leader Evening News,* February 25, 1959)

Our hearts are heavy in this town
Sympathy is all around
For our brothers, who haven't been found,
Mine disaster! We dread the sound!
May God with these loved ones be
Giving strength to them and each family.
We of the town, who truly care,
Our hearts are with them daily in prayer.

Fig. 60. The Susquehanna River frozen at Pittston, much as it was days before the River Slope break-in, circa 1955. (Courtesy of Cooper's Restaurant, Pittston)

### POEM: "MINE DISASTER"
#### BY MARION L. STEFANOVICH OF DUNMORE

(*Times-Leader Evening News,* March 19, 1959)

The morning dawned like other dawns
That fateful wintry day,
The last dawn breaking for twelve men walking—
Unaware of this were they . . .
Off to the job to Shaft Number Nine [May Shaft]
These fated men did go,
With a last look around before they went down
To the dark mine below . . .
Their minds untroubled, voices rumbled
As they went along the way,
But a bit of unrest was in each breast—

Close to the river were they . . .
Not a streak of light from up above
Just battery lamps to guide,
From this sickly light-a trickling wet
Someone's eye espied . . .
From a crack in the wall the water seeped
Soon the crack became a hole,
Then the pregnant river heaved its weight
In an angry, rushing flow.
Seized with fright in their awesome plight
Still they hoped and prayed,
But with all exits blocked, in a trap they were caught
And none could come to their aid . . .
Now all are awakening as another dawn's breaking
As it dawned that tragic day,
But never again for twelve fated men
Who forever in darkness lay . . .

## Musical Composition: "Port Griffith, Pennsylvania"
### Words and music by Frank Murman and Tom Murman

(Recorded by the Murman brothers
on Mask Records, 1959, 45 RPM)

Narrator speaking (with music in the background):
A fateful day in January,
In the year of '59,
The twenty-second we'll remember,
And recall for a long, long time.
In a town in Pennsylvania,
Disaster struck that day,
Bringing sorrow to so many
And we all knelt down to pray.

Verse:
Disaster in the coal mine,
Where good men meet their doom,

Mid fire and explosions,
Families plunged in gloom.
A hundred years no stranger,
To the men who work below,
Port Griffith, Pennsylvania,
Where the Susquehanna flows.

Narrator speaking:
The miners left their home that day,
As they often had,
Thirty-three [69] would return,
And twelve homes would be sad.
The river, it was flooding,
With melting snow and ice.
But no one had a hunch,
That disaster would strike.

Verse:
Disaster, in the coal mine,
On a cold and wintry day.
The river once so peaceful,
Broke through this loathsome way.
The men who climbed to safety,
That some were trapped below.
Port Griffith, Pennsylvania
Where the Susquehanna flows.

Verse:
Disaster in the coal mine,
Where good men meet their doom.
Mid fire and explosion,
Families plunged in gloom.
A hundred years no stranger,
To the men who work below,
Port Griffith, Pennsylvania,
Where the Susquehanna flows.

# Poem: "Second Anniversary"

### by Mrs. Caroline Baloga and children of Port Griffith

(Wilkes Barre *Sunday Independent,* January 22, 1961)

[Mrs. Baloga:]
The twenty-second is today
And we still cry and pray;
The children long for you
And God only knows I do too.
No matter what I say or do
I just can't forget you;
For the things we did and shared alike,
I just don't know if I'm doing what's right.

Fig. 61. The Baloga family receiving assistance from the International Ladies' Garment Workers' Union following the disaster. From left to right: Clementine Lyons of the ILGWU; Sandra, Donald (standing), John Jr., Mrs. Caroline, and Audrey Baloga; and Min L. Matheson, Wyoming Valley District Director of the ILGWU.

When I needed help I called to you
And you always came to my rescue.

But now it's not the same,
Because you're not here when I call your name;
A home that was so happy and true
Is filled with memories of you.
I try to smile but instead I weep
Because you're under ground so deep;
A mine disaster we can't forget
Since your body is down there yet.

[The Children:]
You're needed and wanted, Dear Dad,
Here at home it's very sad;
God took you away from us so fast
And we asked Him why couldn't it last?
I guess God has strange ways of doing things
That we don't know;
And somehow he picked you
And you had to go.

How much you suffered we will never know;
And God gave us a tremendous cross to bear,
For some good reason we must always wear.
God took away the only one we love
But our prayers will go to him up above,
God gave us happiness and took it away
And to Him we will always pray.

Yes we know we will meet him some day,
But the twenty-second is the day
Our Dad from us was taken away;
The year was nineteen fifty-nine,
And for us two years is a long time.

## Poem: "Disaster"
### by Edward A. Goodford of Wilkes-Barre

(*Times-Leader Evening News,* January 1962)

On January the twenty-second—in fifty-nine,
The day shift walked into the mine;
They had no idea that on that date,
Twelve fellow men would meet their fate.

Down in the mine—each miner swung his pick,
Under a roof only eighteen inches thick;
They did not know, that above them lay,
The river, only eighteen inches away.

Dig right here—they were told, not asked,
As the snows above were melting fast;
The river was high, the roof was thin,
The miners never thought it would break in.

They knew it was above them, somewhere,
But no one knew exactly where;
Only time will tell what damage will be done
By the disastrous victory the river has won.

## Musical Compsition: "The Knox Mine Disaster" (Figure 62)
### Words and music by Charles Rogers

(Recorded by The Irish Balladeers on their album,
*The Molly Maguires,* 1968)

Well you heard of the Knox Mine Disaster,
It happened just short years ago.
It occurred in the shafts of a dark, dreary mine
With the land above white with snow.

Well the mine boss said, "Go rob that pillar,"

Fig. 62. The Irish Balladeers. From left to right: Charles (Chuck) Rogers, Todd Spinoze, Robert (Bob) Rogers. (Courtesy of Bob Rogers)

"And you'll do it for the money I love,
And to hell with the safety inspector,
And be damned to the river above."

Now eighty-one men they were workin',
And the diggin' was goin' quite well.
They blasted away that huge column of coal,
And with a great roar, down it fell.

Well the mine roof collapsed at the moment,
And the river poured into the hole.
And eighty-one miners, they ran through the shaft,
Like the evil was after their soul.

Some men made their way to the sunlight,
And thank God, He chose them to stay.
But twelve of their comrades to this very day
Must lie in their watery grave.

Now the mine shafts are deep and they're narrow,
Made greedy men rich in their day,
But everyone knows that the
Black diamond holes
Have taken much more than they gave.

### Poem: "Susquehanna Eternal Rest: Mine Disaster 1959"
#### By George H. Myers of Nanticoke

(*Times-Leader Evening News,* September 1969)

Verse:
    The Susquehanna was filled with gloom
        upon that dreadful day.
    The mine broke through and twelve miners'
        bodies swept away.
    The Susquehanna howl drowned the miners'
        moans of pain.
    Twelve men who loved Susquehanna,
        drowned never to come back again.

Chorus:
    Crushed into dust by mine rock: earth took
        them to her breast.
    And midst the roar of raging Susquehanna,
        gave them eternal rest.
    Somewhere, twelve miners' bodies 'neath
        rock silvered gray;
    Wives and mothers still yearning for their
        miners' return to stay.
    Susquehanna, what hadst thou done? What
        error hadst thou made?
    That along the Valley's river banks such
        terror had been laid?
    The wives and mothers await their miners'
        return—Ah! They can wait in vain.
    Susquehanna eternal rest. Only in eternity
        will they ever meet again.

## Poem: "In the Valley of Coal Below"
## (in memory of Amadeo Pancotti)
### By Rena Baldrica of Plains Township

(*Citizens' Voice,* January 22, 1980)[2]

No words can describe the hurt of heart
Nor time to worry more,
The anthracite and miner part
'tis sealed, by a flooded door.

The miner and his comrades,
Work from morn till night
With pick and shovel flying fast
By oil light, to fill ten cars, the task.

Seven sticks of dynamite they tie,
To blast away black solid wall,
Run: to safety they all ran, a-while
The compound makes shambles of it all.

The roof of coal it trembles, listen!
Of a sudden, down comes a rushing roar.
The Susquehanna covered miners digging
The coal, they want no more.

Mine disasters are tragic and many
In the valley of coal below,
Destiny creates a hero this day,
To safety he guides, two and thirty men.

How sad it is in what we know
Below the surface of this earth,
Twelve miners are still afloat
Not finding ever, their solid berth.

Fig. 63. Amadeo Pancotti, holding the Carnegie Medal. (Courtesy of Pittston *Sunday Dispatch*)

> Forever we search 'neath the ground
> In the valley of coal below,
> Among walls of black diamonds,
> Our twelve miners were never found.
>
> Doomed was the Knox Mine, as all there-in,
> Water, ice-block, rock and debris,
> Fast and furious, water did flow
> Forever, into the valley of coal below.

A miner's life is worth not a damn
In the valley of coal below,
For companies words are but a sham,
Their true value of life is low.

The reason for our living
And time how each and one must die,
'Tis then our sacred soul will sigh
Good-by, they end our time for giving.

### Poem: "Mine Disaster"
#### By Rena Baldricia of Plains Township

(*Citizens' Voice,* January 22, 1983)

One score and four,
   The Knox Mine Disaster
Was permitted
   On the Devil's floor.
This tragic day
   Forever to remember;
Centuries away.
   But God intervened.
Seventy miners he saved
   From a dark water's grave.

From company hands,
   False orders they relay
For miners to dig
   Black diamond coal
No matter how or where.
   Time not be delayed.
For profit was the goal.
   But God intervened,
Seventy miners he saved
   From a dark water's grave.

What heartbreak and folly
   It was to convey
That twelve coal miners'
   Bodies just floated away.
Their comrades alive
   Scrambled and clawed
To reach the light of day.
   But God intervened,
Seventy miners he saved
   From a dark water's grave.

Forever interred
   Twelve souls roam
Haunting each passageway.
   Their buddies they search;
Wherever they may lay.
   But God intervened,
Seventy miners he saved
   From a dark water's grave.

No miner will dig coal,
   For under water is their claim.
The families of all
   Survive with pain
In sadness and prayer
   As they light a candle
Beneath their loved one's name.
   But God intervened,
Seventy miners he saved
   From a dark water's grave.

### Poem: "Knox Disaster, January 22, 1959"
#### By George DeGerolamo of Pittston

(*Citizens' Voice,* January 22, 1986)

January twenty-second nineteen fifty-nine,
Happened the worse disaster at the Knox Coal Mine.
The angry Susquehanna showed its wrath,
Broke through its bottom, destruction lay in its path.
About eighty some miners were in the mine that day,
Never dreaming some would not see another day.

"Everybody out" was the frantic cry.
"Let's get to the carriage or we're all going to die."
The coal miner's tradition to help his brothers;
Many unsung heroes helping the others.

All but twelve struggled to get out in time,
But the twelve still remain down deep in the mine.
Mental anguish and weeping faces on the outside,
Praying their loved ones would come out alive.

The miners toiled many years from early birth,
From pits down deep in the bowels of the earth.
It was coal that helped make this nation of ours,
And sweat of the brow from long working hours.
Thank you kind servants for a job well done,
Your deeds were the greatest equal to none.

To keep on writing would bring me to tears,
To tell all the story would take many years.
So farewell brave miners that lay in the deep,
We know the Lord will watch while you sleep.

Fig. 64. The Donegal Weavers at Eckley Miners' Village. From left to right: Ray Stephens, Emmet Burke, John Dougherty, George Yeager, and Joseph Jones. (Courtesy of Ray Stephens.)

## Musical Composition: "The Ballad of Myron Thomas" (Figure 64)
### Words and music by Ray Stephens

(Recorded by the Donegal Weavers on their album,
*Memories of Pennsylvania's Coal Mines,* 1992)

The morning calm was shattered on that day in fifty-nine.
The whistle blasts meant there had been disaster in the mine.
And blood ran cold for they all knew that lives again were lost.
The price of coal was set in blood and miners bore the cost.

Myron Thomas ran to see what he could do,
For a foreman has a duty to the welfare of his crew;
He found the ashen motorman, who said "I just don't know,
If we can hope to rescue those poor souls trapped down below."

The Susquehanna's broken through, the water's rising fast,
The mine's a churning cauldron, the pockets just won't last!
The timbers are all giving way, the roofs are coming down,

Fig. 65. Mineworker rescued from the Eagle Air Shaft.
(Courtesy of Stephen Lukasik)

Your crew's still in the River Shaft, I fear they'll all be drowned!

Myron rode a car down to the tunnel where his men
Were loading coal that morning, not knowing of or when
He'd see the light of day again, or die down in that hole,
But he must try to free them from that prison made of coal.

In darkness he assembled thirty-one of his lost men,
The water soon claimed seven, who were never seen again,
But twenty-four [twenty-five] stayed with the man they trusted with their lives,
He offered hope they'd see again, their children and their wives.
They made their way back to the shaft where they had all come in,
But rising water claimed it and it looked like it might win,
They prayed the psalm that strengthened and fought for every breath,
As they walked through their valley in the shadow of black death.

They fought back to the Eagle Shaft, closed forty years ago,
It was to have been sealed back then, but owners were quite slow,
To spend their fortunes for such things that brought them no return,
Such safety regulations met with disdainful scorn.

They found an air shaft [cave-in] one foot wide and twenty feet in length,
And thru this void they crawled and clawed and dug with all their strength,
Not one man lost his courage though their chances were but slight,
Then Myron breached the old main shaft "My God. I see some light."

The rescue team soon heard their cries and lowered down a line,
And one by one delivered them out from that dying mine,
Myron was the last one out and turning he did say,
"You'll never claim another life, for this is your last day!"

### Musical Composition: "Stigma"
#### Words and music by Ray Stephens

(Recorded by the Donegal Weavers on their album,
*Memories of Pennsylvania's Coal Mines,* 1992)[3]

The land lies stripped and bloodied, a victim of that time
Great monuments of culm bear mute witness to the crime
Of those accursed villains who raped this land of ours
And left us with this heritage and these unhealing scars.

The veins of coal had been worked out the owners did proclaim
"To hell with regulations, that coal all looks the same
Blast those barrier walls away and take those pillars out
No sense in wasting that good coal, it's safe enough no doubt!"

The northern field was flooded as the weakened roof gave way
The industry that flourished now a thing of yesterday
The plunder took much more than coal, it took a way of life
And all that those mines bring us now is misery and strife.

The era ended suddenly in sorrow and in shame
But years that followed bear the brand of King Coal's brutal reign
The spectre born of baron's greed still haunts the land today
No future for our children here; they grow and move away.

Those endless culm banks blight the land, the tunnels still cave in
Taking homes down with them as the price of all that sin
Whole towns are lost to fires that still rage down below
The aged miners gasp for breath to tell their tales of woe.

There's no doubt that anthracite helped make this nation great
But those who put their profits first have sealed the region's fate
So cursed be those villains who raped this land of ours
And left us with this heritage and these unhealing scars.

### Poem: "The Day the River Came in the Mine"
#### by Eric McKeever, 1999[4]

The miners worked at their humble trade;
　The tunnel so cold, dark, dirty and wet,
Dad and Grandpa had wielded the spade
　Caught in anthracite's shining black net.

Fig. 66. Mine disaster illustration by Edgar McKeever. (Courtesy of Edgar and Eric McKeever)

They knew of the river, so close above
  Swift, quiet, deadly and cold.
Death lay in the wet grip of that glove,
  One touch, and you would not grow old.

Plain modest men with bills to pay,
  A family to raise, it was a hard life.
The dangerous job was the surest way
  to have a family and a wife.

The women at home worked just as hard,
  Cooking, cleaning and sewing all day.
Growing some food for them out in the yard,
  you did for yourself on the small miner's pay.

Beer and music, dance a Saturday night jig,
  On Sunday in church some peace was found.
The rewards of that life were not very big,
  after a week of toil deep underground.

The coal seam passed under the river,
  The old miner spoke with soft pride.
Nature here was a reluctant giver,
  the coal beneath a menacing tide.

"It was good coal, high coal,
  it shone like diamonds, pieces like that!"
Encircling his arms, again in the role
  of a miner, in the chair where he sat.

His encircled arms measured coal that size,
  it loaded the mine car very fast.
The heavy burden of toil might rise,
  be a little lighter, for a spell, at last.

"Around 6 in the evenin' I began to get sick.
        It was black damp comin' into our mine.
We gotta get out, I'm getting awful sick."
        The water was pushin' it into our mine.

"The other guy was real big." I says to him,
        "I'm sick! I can't carry you if I get overcome.
We gotta get outta here soon!"
        Water, forced the blackdamp, making them numb.

The elevator will take them to the top.
        Joe rang the bell, only silence overhead.
No sound of the car beginning to drop,
        If it took too long, they would both be dead.

The gas got heavier, musty, a smell of decay,
        the men stood there helpless, awaiting their fate.
Could the river be coming in, across the way?
        Another few minutes might be just too late.

Joe thought of cousin Frank, removing his hat.
        "He ask me to go wit' 'im," said Joe.
"I told him, 'Naw, I'll stay where I'm at.'"
        The work was the same wherever you'd go.

The Susquehanna gushed with a monstrous flow
        into the mine. Frank would never be found.
A whirlpool of death to those trapped down below,
        he vanished with eleven others underground.

        The crude metal cage at last came down,
                Joe and the other man crawled inside,
        At least they knew they would not drown.
                Joe thought of Frank on the upward ride.

        The Northern Coal Field ended that day,
                for hours the river poured into the mine.

Fig. 67. Nineteenth-century mine disaster illustration. (Courtesy of Chester Kulesa)

By the river, families watched in dismay,
    of life below there was no least sign.
A foreman had worked there since he was a boy,
    he felt the blast of cold wet air.
The cruel flood would crush them like a toy,
    they might survive, the chance was rare.

In a part of the mine, long worked out,
    There was an air shaft, narrow and steep.
Thirty-one miners in a headlong rout,
    The rushing waters claimed seven to keep.

They clawed their way up a narrow slope,
    no more than a foot of space.
The hungry water growling an end to hope,
    their lives the prize in this harsh race.

Twenty-four lived the terror of this night,
    twelve more must live in memory alone.
Before a village church, flowers in bright daylight,

their names engraved on a polished black stone.

Honor to your memory, Cousin Frank!
    You asked for only a place to work.
The work of the miner holds no high rank,
    where you toiled, death was known to lurk.

Now you join the legion of the lost
    who have mined this beautiful anthracite coal.
Energy for the nation at terrible cost,
    Thirty-five thousand lives, a century's toll.

The old miner sat back, he'd finished his tale.
    When the river came in, he was young and strong.
Four decades had passed, what was strong was now frail.
    Joe was pleased that he had lived so long.

After seventeen years a rock fall broke his back,
    removed him from the toil of the mine.
He regarded himself as fortunate in fact,
    to survive, and work another day, was fine.

This plain old man on his porch in the sun
    his wife of his life beside him.
The quiet struggle to survive he had won,
    his pleasure in it would not grow dim.

Silent, I heard the words that he spoke,
    there was nothing more I could add.
I felt honored that I had met such folk,
    joy to him for whatever life he still had.

### MUSICAL COMPOSITION: "THE KNOX COAL MINE DISASTER"
#### WORDS AND MUSIC BY LEX ROMANE (FIGURE 68)

(Recorded by Lex Romane on his compact disc, *Diggin' Dusty Diamonds–Songs from the Coal Mines,* 2002)

It was January back in '59 in a Pennsylvania coal mining town
81 miners standin' in line, waiting to go down
Down, down in that ol' Knox Mine, never see the light of day
Robbing the pillars was the task at hand
Listen what the Bossman say
Listen what the Bossman say

The Susquehanna is a mighty river
It was heavy with ice that day
Right in the middle of a January thaw
That's what the old folks say
Some "rockmen" worked at the River Slope Mine
The great "main vein" was the prize
They kept on diggin' right past the "stop line"
They were tricked by the Company Lies
Tricked by the Company Lies

They kept gettin' closer to the bed of the river
They never really knew where they were
Then POP and CRACK, beams began to splinter
Dust hung in the air
When the river burst in, sounded like thunder
The miners ran to higher ground
But 12 men died in the cold black water
And their bodies were never found
The bodies were never found

The water flowed from mine to mine
All through the Valley it spread
And when it finally came to the end

The whole damn industry was dead
They tried and tried to pump the water out
Oh, but to no avail
Three thousand miners lost their jobs, but only five [6] men went to jail
Only five [6] men went to jail

The head of the Knox Coal Company
Had a secret deal in play
A silent partner, the local boss of the UMW of A
They violated the Taft-Hartley law

Fig. 68. Folk singer Lex Romane. (Courtesy of Lex Romane)

They tried to have it both ways
An inside story of greed and corruption
And the miner pays and pays
And the miner pays and pays

The miners searched for some way out
Joe Stella said, "come follow me"
He led those men to the old Eagle Shaft
A ray of sunlight they could see
Amadeo Pancotti was a real life hero
He scaled up the 50 ft. wall
He found the way out through an old air shaft
The others followed, one and all
The others followed, one and all

A giant void in the mines below
The workers were a-battlin'
A swirling whirling funnel of water
And the river kept a rushin' in
They rolled 60 gondolas into the river
400 other coal cars as well
The whirlpool sucked 'em down one by one
It looked like a railroad to Hell
Looked like a railroad to Hell

So if you're ever in the town of Port Griffith
There's a tombstone there to see
It stands in front of St Joe's Church
A reminder of the tragedy
You know they never went back in that 'ole Knox Mine
They never went back in that hole
It's a watery grave for 12 brave men
Who died for that old Black Coal
They died for that old Black Coal

## Musical Composition: "When the Walls Came Tumbling Down: The Knox Mine Disaster"
### Words and Music by Adrian Mark Bianconi, 2004

When the walls came tumbling down
12 were trapped beneath the ground
we were sad and blue when the river broke through
and the walls came tumbling down.

Near the Susquehanna deep
81 did slowly creep
to their jobs below deep within a hole
near the Susquehanna deep

Though the Knox mine seemed so sound
where the Anthracite was found
men would chip away more and more each day
'cause the Knox mine seemed so sound.

There the icy river flowed
over men who worked below
all that heavy weight opened a gate
where the icy river flowed.

Chorus

They were way down deep
in the chambers far below
when the walls caved in
and the water surely flowed
through the ice and timbers
they waded ever slow
saying prayers, hearing cries, wondering why.

There the miners brave
never giving up on hope
worked the day and night

helping everybody cope
as they paid the price
12 were lost beneath the slope
saying prayers, hearing cries, 12 did die.

Then a sudden burst was heard
no one spoke a single word
all the men took flight in a run for life
when a sudden burst was heard.

One by one they made their way
Climbing out of the wet grave
69 came out leaving 12 in doubt
as one by one they made their way.
In the terrible big hole
many rail cars did go
patching up the leak was quite a feat
it was a terrible big hole.

As the daylight grew so dim
the chance of life grew slim
we began to pray for the souls that day
as the daylight grew so dim.

*Repeat chorus—play and sing—then*

12 were buried in that hole
'cause the owners sold their souls
sealed in a grave are the miners brave
12 were buried in that hole.

When the walls came tumbling down
12 were trapped beneath the ground
we were sad and blue when the river broke through
and the walls came tumbling down.

We were sad and blue when the river broke through
and the walls came tumbling down.

## Notes for Chapter Six

1. The first two poems in this chapter were given to the authors as newspaper clippings by a victim's family. Searches through local newspapers could not uncover the sources. Because of their relevance to the story, however, they are included. Two other poems were provided as clippings with only the month, year, and newspaper (but not the date) listed; they are also included.

2. An addendum following the poem reads, in part, "In Memory of Amadeo Pancotti. . . . For his heroism, he received the V.F.W. award by Memorial Post 6518 of Exeter, the Medal of Honor by the U.M.W [United Mine Workers] of America, also the Carnegie Hero Bronze Medal award." The author could have added two other awards to the list—the Joseph A. Holmes Safety Association's Medal for Honor, and a unanimously adopted resolution from the Pennsylvania House of Representatives citing Pancotti's bravery. These citations make Pancotti the most decorated hero of the disaster. However, his many well-deserved accolades raise a perplexing question: why were not the two other widely recognized heroes, Joe Stella or Myron Thomas, recognized with similar honors? The speculative answer may be that Stella, as a surveyor with the Pennsylvania Coal Company, had discovered the off-course mining that caused the disaster and, although he reported the finding to superiors who did not act upon the information, award agencies may have wanted to avoid possible controversy. With regard to Mr. Thomas, his managerial position as assistant foreman with a company that had mined in illegal territory may likewise have tarnished his candidacy in the eyes of evaluators.

3. The Donegal Weavers included the following preface to the words of "Stigma" in the booklet accompanying the album. "Although the Northeastern Pennsylvania anthracite era helped forge the industrial might of the United States, it left in its aftermath a dark legacy. 'Stigma' portrays the bitter memories of a culture which fell victim to the lust of the 'mine barons.' The song metaphorically compares King Coal's brutal reign to a rape which left the region ravaged and despoiled."

4. In this poem Eric McKeever writes about Joe Brady, a Knox survivor who lived next door to McKeever's in-laws' home in Pittston. Brady talked to McKeever about the loss of his cousin, Frank Orlowski, at Knox. Originally written by the author to commemorate the fortieth anniversary of the disaster, the poem also appears on the Web site of the *Anthracite History Journal*.

# Chapter Seven
## Memorializing and Relecting on the Knox Mine Disaster and Anthracite History

*... the suddenness with which this disaster came to our community has utterly shocked and crushed us.*
George DeGerolamo, retired mineworker

*These places [need to be] saved and the history shared with generations to come.*
William Best, president, Huber Breaker Preservation Society

*... if they [youth] had a little more grounding in it [their past], if they knew their genealogy, if they knew their grandparents, or great grandparents, or great-great [grandparents]—the ones that really had the hardship to get here—and did a good job—they would be better off.*
John Dziak, co-founder, Greater Pittston Historical Society

Fig. 69. George DeGerolamo (holding paper) reads eulogy at the May Shaft on the first anniversary of the Knox Mine Disaster, January 22, 1960. (Courtesy of Stephen Lukasik)

## Eulogy: "Hush in Our Hearts—To the Twelve Men Who Lost Their Lives in the Knox Disaster"

### by George DeGerolamo of Pittston

(Delivered at the May Shaft on the first anniversary of the Knox Mine Disaster, January 22, 1960. From George DeGerolamo, *Autobiography*, published by the author, Pittston, Pa., 1980)

My Friends, this morning there is a hush in our hearts as we come together to pay our respects to the memory of those twelve men whom so recently were well and happy, but who now have passed into the great beyond.

There isn't anything unusual about death. From the very beginning of time, people have seen their loved ones pass from life to life beyond. But the suddenness with which this disaster came to our community has utterly shocked and crushed us. In time of war we expect many to be killed, and we resign ourselves to whatever outcome there may be. In time of peace when plagues and pestilence sweep out thousands throughout the land, we prepare our minds for the tragedy which may visit us any day, and thus we are spared the stunning repercussions. Had it not been for greed and financial gain, this would not have been. For "What profit a man if he gained the whole world and lose his soul?"[1]

## Oral History: Sam DeAlba

In 1982, Sam DeAlba (figure 70) of Duryea became one of the driving forces behind The Knox Mine Disaster Memorial Committee. The group led to effort to place a monument in front of St. Joseph's Catholic Church in Port Griffith listing the names of the victims. The black granite marker was dedicated on January 23, 1983, for the twenty-fourth anniversary. The committee also participated in securing a historical marker from the Pennsylvania Historical and Museum Commission in 1999. His (and his fellow committee members') remarkable dedication to these commemorative tasks provide a clear indication of the community's con-

Fig. 70. Sam DeAlba, co-chair, Knox Mine Disaster Memorial Committee.

cern for the Knox disaster as well as the area's industrial history. (From Sam DeAlba, taped interview, NPOLHP, July 24, 2004)

SAM DEALBA: I was a member of the Greater Pittston Jaycees, which is a community group. The local newspaper, the Pittston *Sunday Dispatch,* every year runs a story on the [Knox] anniversary, and their remark [in 1982] was [that] there is no marker anywhere in the area in honor of this. It said if anyone wants to get a committee together, call the *Dispatch.* This was on a Sunday. Monday or Tuesday I called the *Dispatch,* and so did Pep Orlando. We got in touch with each other, advertised a meeting, got a hall in Port Griffith, and we had about six or seven people on the committee.

Once the word got around, some family members came [to our meetings]. We said, "Let's do something for the Knox mine [victims]." What we were gonna do, we didn't know. We brought in John Marino from Dupont Monument and we threw around some ideas. His one idea was [a monument of] black granite, which resembles coal. We said, "Okay, now we need to know some cost factors." So he went back to his drawing board [and] he started drawing things up to give us some idea. We sat down, [and said to ourselves], "Now what do we do, what's our next step?"

So we started. We had to start raising money. We got Chamber of Commerce listings of all [local] businesses. We sat down for a month and wrote letters to every business in the greater Pittston area.

This would be the bigger business aspect of it. Get some money in. We also put ads in the paper. Five dollars, ten dollars, two dollars—these started coming in. We got the local Key Clubs involved. We had bucket drops. We had gifts donated. We had drawings, dollar chance drawings. When the money start rolling in we called Mr. Marino back and he gave us some drawings.

This was six or eight months down the line. It was time consuming, very time consuming. After he came up with the drawing we got in touch with some of the family members to come in and look at it ... We had half the money and we start getting worried. Where were we going get the rest?

We put more ads in the paper. Now bigger businesses and a lot of local unions and fire departments and boroughs and townships— even the officials from these townships—all came forward. . . . Exeter,

West Pittston, Hughestown [many municipal governments]—they all made some kind of donation. A few hundred dollars. . . . We did meet at a city council meeting and made a presentation. That's how you got the other groups involved. They'd see the picture in the paper and, "Oh, I got Pittston Township to donate the other day, can you do it?" They saw the picture in the paper then they start calling me. [They'd say] "Well, everyone else was doing it, lets donate $50, let's donate $100," and it was just like a snowball going downhill.

ROBERT WOLENSKY: So you used a lot of publicity.

DEALBA: Yes, a lot of publicity. Every other week our picture was in the paper. People were sick of me. (laughing) I pushed that a lot, [in] every group I belong to. You have to let the people know what you are doing. . . .

A lot of the unions came through. The coal miners' union, the AFL-CIO. A lot of these people came through with nice donations. A lot of bigger businesses came through. Not substantial, but enough. You get your $100 off twenty businesses. But most of the money came from ma and pa—the little guy, small donations. Fundraisers were hard. We made money on fundraisers. We did chances and had all the prizes donated, coin drops at intersections on local roads. Most of the money came from the little person. [It took] over-about a year-and-a-half to raise all the money, but it was worth it. It was well worth it. We were committed. I had a good committee. It was hard but. . .we met and surpassed our goal. . . .

Then we said, "We have the money. Let's order this." Mr. Marino came back, showed us the finalized drawing, and everyone approved it. It was a long ordeal, it was, but it was very worthwhile. . . .

WOLENSKY: You were probably the man to do this, in that sense that you knew where to go.

DEALBA: I know where to go. I knew where to go to get money off people (laughing). You don't beg, you just make them feel like they have to do it. [It's] their responsibility to the community. Those are their customers that you're helping out. Scratch each other's back. It's hard asking for donations all the time but sometimes you get blessed at the end. . . .

WOLENSKY: Who was on the committee, Sam?

DEALBA: Pep Orlando, myself, Alfreda Suchocki, Jean Talipan, Michael

Cotter, Jean Mangan, and Nora Murtha. . . . (figure 71 and figure 72)

WOLENSKY: What was the final cost?

DEALBA: Approximately $7,000 just for the monument itself. Money was put away for perpetual care and the reason we picked Port Griffith and St. Joseph's Roman Catholic Church [was] that since the day the disaster happened, the priest at this church always had a mass on the anniversary. He was one of the first priests at the River Slope the day it happened. He was at the hospital with all the injured that day.

I forget his name [Fr. Edmund Langdon]. They called him the priest from heaven. "Pennies from Heaven" they called that church because he was always collecting a lot of pennies to build it. The church is located a couple hundred yards from the River Slope, and that's why the monument is on the church's front lawn. Then, after this we did get a state [historical] marker.

WOLENSKY: Tell us about the state marker.

DEALBA: The committee decided this should be a historical site. The Knox ended deep coal mining in Northeastern Pennsylvania [Wyoming Valley] and just about everything else at that time. We got applications from the local historical society and our state historical commission, and we kept getting turned down. We just got disgusted and we put everything on the backburner. Then we went through [State] Senator [Raphael] Musto's office with Mike Cotter [Musto's assistant], who was a member of our committee. We got another paper [application] together [but] we never heard anything.

About ten or twelve years later we were contacted by Ken Wolensky from the state Historical [and Museum] Commission asking us how did we go about getting it? I explained to him that we've tried more than once. He got back in touch with me two weeks later. Everything was on the books; we had everything done. A year later we had our state marker next to our monument in Port Griffith at St. Joseph's Church!

That was another expense [providing local matching funds for the state marker] but we always had this in the back of our minds. We did continue our fundraisers for another six months so we had money for the monument, the marker, and some kind of perpetual care—flowers, landscaping. Now it's over twenty years later and we're still keeping it up. . . . It adds up after twenty-five years, the expenses.

Fig. 71. Knox Mine Disaster memorial unveiled at St. Joseph's Catholic Church, Port Griffith, Pennsylvania, January 24, 1983. Committee Members, from left to right: John Marino (designer), Nora Murtha, Jean Mangan, Pep Orlando, Sam DeAlba, Rep. Tom Tigue, Jean Talpian, and Alfreda Suchocki

WOLENSKY: Who's charged with the upkeep?
DEALBA: Jean Talipan and myself do most of the landscaping and flowers and make sure everything's okay for the holidays, for the anniversary. We remind the priest. Sometimes parish priests change their assignments [so] we remind the parish priest a month ahead of time, "Don't forget the mass." You know, little things like that.
WOLENSKY: Is the fund holding up?
DEALBA: The fund is just about holding up. A lot of things are artificial anymore. You don't put too much live stuff in. The anniversary is in the middle of January [so] you're not [going to use] live things, but we have a nice three-foot wreath that hangs right on the monument.
WOLENSKY: Do you remember the state marker dedication for the fortieth anniversary?
DEALBA: Yes, yes. It was beautiful. Everyone was there. State representatives honored our group for doing this. The local Boy Scout troop—I was an eagle scout in that troop—was there for the ceremony. It took so long but the plaque [marker] and the monument were done

and I was very happy. I was happy for our committee and for the families of these twelve men who died and all of the coal miners of the area. They're finally recognized. Everyone knows where this happened.

WOLENSKY: How many years from when you first thought about having the monument to getting the monument and the marker?

DEALBA: Eighty-one [1982] we started working on it, so a good eighteen years. . . .

WOLENSKY: How many years before you tried to get the state marker up?

DEALBA: Oh, had to be a good ten years after that. That was the second time we applied. Then about ninety-eight your brother [Ken Wolensky] stepped in and we did all the paper work and everything for that.

WOLENSKY: And you finally got it in ninety-nine, for the fortieth anniversary.

DEALBA: Fortieth. That worked out beautifully. Before we know it, it will be the fiftieth (laughing).

WOLENSKY: We just celebrated the forty-fifth. You were at that mass. I saw you there. In fact, we both marched [with family members] to the front of the church and lit the candles. . . .

DEALBA: Yes. . . .

WOLENSKY: Any future plans with fundraising or anything else?

DEALBA: The only thing we're looking for now is the fiftieth anniversary. Two or three of us see each other occasionally. We want to do something besides the mass, maybe a social event, even if it's breakfast in the church hall. Make sure that the Lackawanna Coal Mining Association [Anthracite Heritage Museum] up there knows we picked the right date, so we can work in conjunction again. And the new [Greater] Pittston Historical Society, maybe we can do something together, make it a whole weekend like they did on the fortieth. The church was full [for that event], and in this day and age most churches are not full any more!

WOLENSKY: I recall Judge [Gifford] Cappellini being there and [State Rep. Thomas M.] Tigue.

DEALBA: Tigue and [State Rep.] Raphael Musto, and [House Rep. Paul E.] Kanjorski. The locals were there, mayors. Mayor [Michael

Lombardo] of Pittston spoke. Most of them come out for this, not every year but when it's a big anniversary these guys, they're busy, but the try to make it. . . . .

WOLENSKY: Were there any other reasons for your active involvement on the committee?

DEALBA: I got involved not [mainly] because of the Jaycees, but both [of my] grandfathers worked in the mines. My one grandfather died of black lung in his sixties. My [other] grandfather on my mother's side died at age thirty-eight, leaving the whole family behind. He died from an injury in the mine. I never met this grandfather and my mom and her brothers and sisters just about remember him. He died as a young man, left my grandmother alone with the children. And that's in my heart. That's the reason I got involved with the Knox mine [memorials]. Neither one worked at the Knox but they both worked in mines in the greater Pittston area.

WOLENSKY: Were your grandparents born in Europe?

DEALBA: Yes, they were born in Italy. There was one from Sicily and one from Calabria.

WOLENSKY: So your heritage was in the back of your mind.

DEALBA: That was [why] I was so active at the time. It just got to the point [that] I wanted this monument and I wanted this monument and I wanted this monument! It worked out. It took us a year and a half but it worked out beautifully.

WOLENSKY: I wonder whether the area's working class history motivates you?

DEALBA: Yes, it does. My parents. My dad worked in the garment industry and so did my mom. My dad worked for the main pants factory for forty-five years in West Pittston. He was in the garment industry his whole life. My mom worked dress factories here and there. Those were the jobs around here. That's all there was around here in the fifties and the sixties. My parents were hard workers. They never made a lot of money but we always had what we needed. We had food on the table and a roof over our head. And children today, they appreciate it but they don't come right out and say that.

WOLENSKY: It has been said that youth today often don't really know the depth of the past. Do you think that's why the Knox commemorations are important, so people know who they are and where they come from?

DEALBA: Yes. My kids know I'm involved because I care. I'm always doing something for this. Yes. . . .

WOLENSKY: Are your children interested in local history?

DEALBA: My daughter more [than my son] because she was there when we were doing this in the eighties. My daughter reminds me, "When is that mass dad? If I'm in from school, I want to come to the mass." We try to attend the mass every year. . . .

WOLENSKY: Some people would ask you, "Why are you still commemorating this disaster?"

DEALBA: Not just the Catholic religion, but any religion around here, it's embedded in you. History is embedded in you. The coal mines—you can still see the scars around the area. You can still see culm banks in the area. You still hear about [underground] coal fires. People [are] back using coal in their homes. It's just a circle. Even just picking up the newspaper two weeks before [the anniversary], the local *Dispatch* puts in a nice spread. The [other] papers put a nice spread in, and then it sticks in your mind about this Knox mine [disaster] and the coal industry. It just sticks there

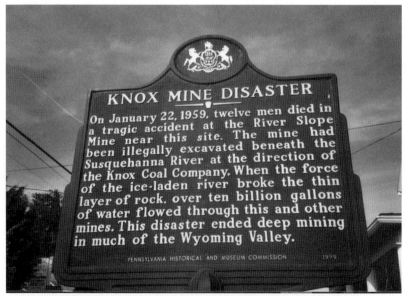

Fig. 72. Knox Mine Disaster state historical marker placed at St. Joseph's Catholic Church, Port Griffith, Pennsylvania by the Pennsylvania Historical and Museum Commission, January 23, 1999.

Fig. 73. William Best, West Pittston, preparing the marker at the site of the River Slope break-in, July 2003.

When you drive down Main Street in Port Griffith—thousands of cars [pass there every week]—and the marker and the monument are staring you in the face. At the other end of Pittston there's a coal mining statute, which was put up in the mid-seventies. You drive down Main Street Pittston, that monument is there, a full miner. The Lackawanna Coal Mining Tour is just up the road. It's just—it's our history, in this greater Wilkes-Barre/Scranton area. Coal mining. Most of our grandparents worked in the coal mines. And then you pick up the newspaper [and] there are disasters all over [the world] still going on today—people dying, people being saved in coal mines. It's a job that's dangerous, but it's a good paying job and young men and older men go down in those mines. They don't know if they're coming back up—just like in the fifties.

## Oral History: William Best

William Best (figure 73), West Pittston, current president of the Huber Breaker Preservation Society, is one of the area's leading preservationists.

During the fall of 2002, he joined with colleagues in finding the exact spot of the Knox breach, as well as the location of the Eagle Air Shaft. Through his efforts a marker was placed on the site of the breach in 2003 (figure 74). He discusses the marker, the meaning of the site, and his future hopes for it. (From William Best, taped interview, August 9, 2004, NPOLHP)

ROBERT WOLENSKY: Bill, would you begin by telling us what gave you the idea for the marker at the site of the Knox break-in?
WILLIAM BEST: Well, we started going down there—Joe Stella, John Dziak, and Bill Hastie [and myself]. With the use of Joe's maps we determined how many feet the break-in was from the service buildings on those maps. We used a scale to figure out the distance and then we used a wheel that was marked by the foot to determine how far it was from the buildings. We took the measurements down the [railroad] tracks to the area of the break-in, eighty feet from that point, to determine how far into the woods on the other side of the tracks the break-in occurred. [As further evidence] when we got in there we saw one sink hole [earth settlement] where they filled in [the actual hole] many years ago.

Then we went on down the tracks further and we found the exact spot where the Eagle Air Shaft was [located], from the starting point of the canal building which was part of Irv Griglock's property there. . . . We were able to determine [the Eagle Air Shaft location] because there was an angle iron upright on each side of the opening. Keep in mind the opening was blown shut and there was some fence there, rusted fence, and there was a big old eye hook on the right hand side. We determined without a doubt [it was] the Eagle Air Shaft.

Now when we came to memorializing [the break-in spot], we first went in there and cleared that area [from tree and brush growth]. We had a marble foot marker donated to us by the past director of Veteran's Affairs [in Luzerne County], Art Bartolai. He had some markers that were unusable [because of errors] and what we did was we flipped this marker over on the back side and had Jack Marino from Dupont Monument inscribe the marker. You know, I asked you if would mind giving us some options for how to word it.
WOLENSKY: Yes.

Fig. 74. Marker at the Site of the River Slope Break-in, July 2003.

BEST: And I believe we came out with "The Knox Mine Disaster Occurred on this Spot, Jan. 22, 1959; Twelve Men Died." So the idea was a very collective [thing]. A lot of people put a lot of different effort into the memorial, the wording, the location, the finding. John Dziak videotaped the actual process of finding the spots and scaling [it] out and then [the] further measurements on the day we actually determined [the location]. I have photos, too.

It was probably in the fall [of 2002 when] we found it [the location] and then in early 2003 we started clearing the area and getting the stone inscribed by Jack Marino. I know I dropped it off at the end of 2002 and it was finished by Christmas. We were able to break ground after the frost had cleared up that spring [2003], so that was when we actually planted it.

WOLENSKY: Bill, would you describe how you actually had it set up there with the coal and flags?

BEST: As you walk down to where we determined [the location], we found a flat spot that was near the depression and we dug it out and we placed the foot marker [which] is one by two feet. I'd say that the concrete base that was made for fixing a bronze marker is two by three. It's a little bit wider. What we did was we set that in place and, of course, we dug it out. Incidentally, that was all mine rock there, so we dug down to the topsoil until we hit mine rock and then leveled that out and placed the concrete base down. Then we drove the steel reinforced bars down in the four holes in the base and we put concrete over the top. I made a form of two by sixes to bring the stone up a little more. We mixed the cement and we filled the form up with concrete. Then we laid the foot marker inside the form so that it would be even. Joe [Stella] came up with the idea of getting steamer coal, which is ample size coal, and placed it around the foot marker.

WOLENSKY: And the flags?

BEST: Oh, the [American] flags were set around [for the dedication]. Twelve flags were [placed] symbolizing the twelve men who died in the disaster. Twelve American Memorial flags they were. . . .

WOLENSKY: Who attended the dedication of the site marker?

BEST: That would have been Joe Stella, Bill Hastie, Matty Amico, you, and Father Hugh McGroarty who said a few words [prayers]. He had some holy water [and], blessed the site, and he said some words [prayers]. We have that video taped. John Dziak was there as well. . . .

WOLENSKY: Bill, more philosophically, what was your motivation?

BEST: Well, ever since I met Joe Stella and Bill Hastie, I've given a lot of historical significance to the Knox. I felt very, very strongly about putting up [a marker] to pay tribute to the men who died. [I felt honored to be] in a position to be amongst these great mining men and these great workers and surveyors. Also the significance of the twelve who died coincided with the end of deep mining in our area. By the time I was about old enough to understand, mining was pretty much gone from this area. Only the sites with a few remaining coal breakers, mine openings, and these miners [remained].

I have to say, Bill Hastie and I became very good friends after I heard him speak at the Plymouth Historical Society's 130th anniversary five years ago this September. Prior to that I was very much involved in my schooling, my engineering work, and work in industry. But Bill and I became good friends [and we attended] the ninety-nine [Knox fortieth] anniversary, the 130th Avondale [mine disaster anniversary], and then [we went to] the Huber Breaker [Preservation

Society] meetings. And, of course [there's] Joe Stella, who I have been great buddies with ever since the first time I met him [in 1999]. . . .
WOLENSKY: Bill do you have any plans to make that new marker seen by passersby?
BEST: Well, I think ultimately what I would like to do is to [have it become part of a] walking trails project in that area. I'd [also] like to place a similar marker or plaque at the site of the Eagle Air Shaft with words signifying its significance.

I think ultimately what I would like to do is acquire the land right next to the houses built on the side of the [former Knox] coal pockets and possibly make some kind of a stairway that goes down to the railroad tracks, so there would be easement to that area without having to walk all the way [around] down to the tracks. In other words, maybe a stairs that went down to the tracks and then parking or something at the end of the road, so that any future memorial [would be easily accessible]. Hopefully by the fiftieth [anniversary in 2009] we can do this.

We could actually put a memorial park up there. That would be my ultimate goal. . . . You can raise money to buy things and [in] five years develop it to become a big part of any kind of walking trail. . . In the future [we could] become part of any kind of group that is interested in tying these areas together for the sake of historical restoration. I mean, for me that's my dream—to save this piece of land so it isn't turned into a trailer park or a dump or [used for] something other than [for] historical purposes.
WOLENSKY: Sounds like you see this land as almost sacred turf.
BEST: Absolutely. There's no doubt about it that.... These places [need to be] saved and the history shared with generations to come.... I can never emphasize enough the fact that Luzerne County is really lagging behind Lackawanna County in so many ways. It does not have any significant tribute to the coal miners besides the Luzerne County Historical Society's basement display on coal mining [and] the state historical markers. [These] are the only things that will give you any idea that there was any kind of disaster or mine workings or colliery sites. . . . The Huber [breaker] is definitely one of those sites that is very close to being realized. It's not surely a lock or hasn't been acquired yet but it stands the best chance of [being saved]. . . . You see

Concrete City [a planned workers' housing development]—there's another example of an architecturally [and] historically [significant place]. [It was part of the ] Truesdale Colliery [in Nanticoke].

I think that all of these former colliery grounds are worthy of some kind of a memorial. At least [we can] have the land turned into a park or not developed—to be kept as landmarks. [When you] go from Forest City to Shickshinny, all of these towns, pretty much [exist] because of their location [near] mine openings and collieries. This is what made this area [different] from any other kind of farmland. You know, it would be normally farmland. Why would anybody want to put Exeter next to Wyoming, next to Swoyersville and Kingston, and even more so when you get into the Lackawanna County? All of those towns in the mid-valley beyond Scranton—they all sprung up because of their location to the mines. . . They didn't need to be populated for any other reason than their location [near] the mine openings.

## ORAL HISTORY: JOHN DZIAK

John Dziak of Pittston (figure 75) is a historic preservationist and one of the founders of the Greater Pittston Historical Society (GPHS). He discusses his personal efforts and that of the GPHS to conserve and educate about the Knox mine disaster and other local history sites and events. (From John Dziak, taped interview, August 10, 2004, NPOLHP)

JOHN DZIAK: As soon as we announced our association [GPHS] and put it in the newspaper, we did a program. That would have been April of 2003. The people from the [former] Pittston Hospital [now a office center] called up and said they were very interested in history. So we went down and talked to them. We were able to develop a Pittston Hospital program [involving] former employees. The hospital had been shut down since the early 1980s but there was still a very viable group that was interested since they loved the place. So we were able to get everybody together and provide information about the hospital and we videotaped it. I had enough of my own equipment to do that.

As we were going through it and asking ourselves questions about

Fig. 75. John Dziak, Pittston, videotaping the marker dedication at the site of the River Slope Break-in, July 2003. (Courtesy of William Best

the hospital we said, "Well, what was the hospital involved in, through the years from 1892?" The one thing we tried to find [out] was the relationship between the Knox mine disaster and the Pittston Hospital. The only thing we were able to find easily and quickly was a videotape put out by WVIA. I sat with that videotape for about three hour, trying to figure out how my computer software could take the approximately forty-five seconds of Pittston Hospital [viewing] time and make something out of it. Then I said [to myself], "This is ridiculous, there's more to the Knox and the Pittston [Hospital] relationship than forty-five seconds!" So we started asking nurses about it [during the session], those that were there at the time and those that heard about it. We video taped their comments.

Then something [else] happened. Billy Best [had something to do with it]. Billy is the instigator of many things! You're also one of the instigators! But anyway he said, "Well, why don't we do something on the Knox?" I think he contacted you and you said, "Oh that's a great idea, let's have a [roundtable] discussion session with nurses and others."

ROBERT WOLENSKY: We had the session in May of 2003. . . .

DZIAK: The Knox happened when I was fourteen and it did impact me a bit. Then I immediately forgot it for the next thirty-six years. . . . When the opportunity arose, it was, "Ok, let's learn about the Knox," and then we had this [roundtable] program. I'm sitting there with my two or three video cameras and Billy's sitting there with his camera, and I had an opportunity to listen to [the] people [six panelists] there. And this is what got to me! We were forty-five [forty-four] years later but the people were talking about it as if it were yesterday. And even though we had a full evening and a great discussion, there were more questions that I had for myself than were answered. . . .

WOLENSKY: John, tell us about some of the maps and other items you have made to preserve the physical information about the Knox disaster.

DZIAK: I am visual person. I can't conceptualize something, especially something I am not aware of. . . . Billy got Joe Stella to loan Billy his [Knox mine] maps. Billy went down to Wilkes-Barre and had his maps scanned and got them copied and brought them over to me. I'm looking at them and I had no idea what was going on. So I decided to ask Joe Stella. When he started pointing and telling me stories [of] what happened, I said, "Okay, we're going to get the camera out and we're going up to the [Knox disaster exhibit at the] Anthracite Museum [in Scranton]. We had two cameras rolling. Billy on one side of me, Joe on the other, and Joe and Bill [were] talking. We walked around the whole area with the cameras rolling [until] we ran out of film. We came back and said, "Wow, there's a story here. I still don't [fully] understand it but there's a story here."

Then we [GPHS] organized the [forty-fifth Knox anniversary] program on January 22, 2004. For several weeks before that we prepared one particular [mine] map just so that we could hear Joe tell the story one more time. Every time he told the story we were getting another piece of the information.

The one thing that really intrigued me was at the [roundtable] discussion session [in May 2003]. Don Baloga told a story about his father's clock [that ran inexplicably for several days], and it was something that bothered me.[2] This guy [victim John Baloga] could have been in the mine for ten days. There could have been an air hole [pocket]. I'm thinking about all these things that could have happened, you know—the ground could go up, there could be a place where he could survive. I had the maps and I started looking at the contour maps and the topos [topographical maps] and I thought, "I've got to do something." I put it [the maps] into the computer and I played around until I found [some answers]. In my estimation, and I'm just guessing from what I could see visually, that they [most victims] were gone [drowned] within the first half-hour. For the people who were in the lower veins [it might have been a little longer]. But anybody who was in the Pittston and the Marcy [Veins] were probably gone within the first thirty minutes. Of course, the first three [the rockmen] were gone within seconds.

And the way you could do [judge] that is when you get a topo [topographical] map you could see that the break-in point was at a particular [angle]. You have contours going down the hill [underground] and when you look at it on a map, you start to understand. I actually got to create the map and see it and then, as I created that map, more stories came through. . . .

The [Knox] story is addictive. No sex [but] its got intrigue, its got emotion, its got violence! We just have to figure out how to get sex into it and we'll have it made! This would be a natural for television. And just the way it's told in the Pittston vernacular, it's a story that has to be told. [Yet] there has to be more than me, and you, and Billy, and Joe that would understand it, because when we go [die] nobody will. That's what happened with the Twin Shaft [mine disaster of 1896]. Nobody even realizes that there are ninety-eight people buried [bodies never recovered] under the ground a particular spot [in Pittston]. I never did either for the first fifty years of my life.

WOLENSKY: Not to far from your home.

DZIAK: Right across the street. Actually, the shaft is, I'm gonna say, within 250 feet [of my home]. I never knew it was there. Never knew about the disaster. Even after they put the sign up [state historical marker], it was just far over my head. People just don't know about it and I think it has to be researched, the spot has to be noted. People have to realize that some people are underneath there. And, it [information] really has to be put out. Now [the] Avondale [disaster of 1869] is another story and that's being handled by Joe [Keating of Plymouth]. I plan on doing as much as I possibly can with Joe to get that [site preserved].

WOLENSKY: Can I ask you about the paper model you have made of the actual River Slope area?

DZIAK: Okay, I took the map, digitized it, digitized and filled in the contours [and] the [underground] roadways, headings, and gangways that were noted on the map, and put all that in different layers on the computer so that I could strip off each layer and see things. Then I printed out what I wanted, which was basically the contour lines and what was happening on the contour lines. Then I cut out the contour lines and built the [paperboard] model.

WOLENSKY: So you glued the mine layers one on top of the other?

DZIAK: Yes. First time around I didn't have any idea what I was doing and it took me about a week. I never took courses in 3-D modeling.

WOLENSKY: And you have another model that involved a similar exercise?

DZIAK: Well, that one was the first one of the entire area. It covers from a little beyond the Eagle Air Shaft to a little past the [off-course] tunnels area, maybe a couple hundred feet past where the tunnels would be. That gave a broad view of what an anticlinal actually looked like. Then you can see the break-in point and the whole story about what happened [during] the weeks before and how and why they did the things that they did. What we [also] did was to take a portion of the map and blow it up, probably about four or five times bigger than the original, same contour lines, same information, just a smaller portion of the map made bigger. We could see a lot easier. Certainly not life-size or anything, the figures are less than two feet long [and] twenty inches wide. But it shows [everything] when you look at it, and it's a 3-D model. It shows where the slope entered the Pittston Vein—about 420 feet above sea level. [It shows] how close it was to the anticlinal underneath the tracks, and where the tracks were. It shows everything.... You know, like the off-course chambers and the jalopies and all these other big pieces of equipment that might have been in there. The rockmen—where the rockmen [were], visualizing their story. Again, that really helped me visualize it more than reading [about it]....

WOLENSKY: John, in a more philosophical sense, what really drives you in your historical endeavors?

DZIAK: Well, number one, that I am able to. I have the time and I have the equipment... I've always wanted to retire and in order to retire I have to stay out of my wife's way, so I have to be busy at all times!

WOLENSKY: That's a good reason!...

DZIAK: [But there is] another reason that we haven't discussed yet. I believe that what I have collected and what I do should stay for a while, but somebody should understand it, somebody should know [about] it.... [But] you have to make your audience. Britney Spears, Madonna, and those people—it's very easy for them. They just, you know, shake their stuff. We have a more difficult [time] in that people aren't as interested in history and we have to work a little harder gathering an audience.

WOLENSKY: Do you think that young people might need to know more about history and less about than the typical fare the popular culture provides?

DZIAK: They have nothing else. I grew up with a mother and father, a traditional family, and my father let me know it if I didn't do things right. And I had grandparents and I knew them all. But I didn't pay particular attention to anybody. Kids nowadays really...[I don't know]. My background is as a pharmacist and as a pharmacist in the sixties I knew a little bit about LSD and heroin and things like that but I never saw it. It was rampant because people don't understand what

Fig. 76. Knox Coal Company employees, circa 1952. (Courtesy of William Hastie)

it's like to have a natural high. They want to be happy [but] they don't understand what's good about doing something commendable about something or for somebody. They just don't understand it. So if they had a little more grounding in it [their past], if they knew their genealogy, if they knew their grandparents, or great grandparents, or great-great [grandparents]—the ones that really had the hardship to get here [coal region], and did a good job—they would be better off.
WOLENSKY: So you think that such knowledge might provide some food for the soul?
DZIAK: Yes! Yes! God knows, they need it.

### Essay: "Reflections on the Forty-fifth Anniversary of the Knox Mine Disaster"
#### By Robert P. Wolensky

(Slightly different versions of this essay appeared in the *Citizens' Voice*, January 22, 2004, and in the *Anthracite History Journal,* Spring 2004)

I read with great interest the list of commemorative activities associated with the forty-fifth anniversary of the Knox disaster. They included a candlelight ceremony on January 22, 2004, near the disaster site in Port Griffith, followed by the opening of an exhibit, a slide presentation, and an audience discussion at the old Pittston Hospital—events sponsored by the Greater Pittston Historical Society (GPHS). On January 24, the Anthracite Heritage Museum in Scranton organized an anniversary program that included photos and videos, two presentations (one by me and the other by Erika Funke of WVIA-TV), comments by Joe Stella and some victims' family members, as well as a performance by Lex Romane who sang his new song about the disaster.[3] Father Bendick at St. Joseph's Church in Port Griffith celebrated the annual memorial mass on January 25 with victims' family members and survivors in attendance. And, as with all previous anniversaries, the newspapers and other area media provided considerable coverage (figure 77).

While the national interest in Knox soon diminished in 1959, the disaster has continued to receive wide local attention over the past forty-five years. Consider the following list of public events associated with the tragedy: the commemorative marker set by citizens of the Pittston area in front of St. Joseph's Church in 1983; a state historical marker from the Pennsylvania Historical and Museum Commission in front of the same church dedicated on January 23, 1999; a panel discussion on the disaster at Luzerne County Community College in 1996; a fortieth anniversary commemoration program organized by the Anthracite Heritage Museum in Scranton on January 22, 1999, that included an exhibit that ran for three years due to exceptional public interest; the memorializing marker set on the actual site of the breach by members of the GPHS in July 2003; a well-attended community forum on the disaster organized in May 2003 by the GPHS; and the regular airing of WVIA-TV's half-hour Knox documentary.[4]

Why have the citizens of the northern anthracite area so faithfully commemorated the disaster on a regular basis? Indeed, why do the remembrances seem to draw more attention with every passing year? Why can't we forget mining catastrophes such as Knox—or forget the entire anthracite era, as some have suggested—and get on with the future?

I would like to suggest two reasons why this catastrophe has stayed with the community in such an impressively enduring way. The first is that it involved such an egregious violation of the public trust in the coal companies, the mineworkers' union, and governmental regulatory agencies. Numerous other mining calamities had struck the area before 1959, and caving has remained a persistent problem for decades, but few hazards were as upsetting as Knox because of the illegal mining and far-reaching "culture of corruption" that caused it. The event precipitated four government investigations that led to indictments against twenty-two individuals and four coal companies on charges that ranged from conspiracy, to tax evasion, to bank fraud, to labor and mining law violations. Twelve persons, including seven affiliated with Knox, and three companies were found guilty—although, in the final analysis, no individual or company was punished for illegal mining or for negligence in the deaths of twelve mineworkers.

Second, the Knox disaster has become much more than a heartbreaking cataclysm. I believe that it stands as a symbol of the exploitative mining and employment practices of an entire industry. It reminds us of a very dangerous industry that took the lives of nearly thirty-five thousand of our relatives and neighbors. It represents the hard-working men and boys who endured hardship because they wanted to make a better life for their families. In other words, the disaster is still with us because the people and the institutions of the area will simply not allow it to be forgotten.

Therefore, while the immediate effects of the disaster were emotional and economic, I believe that its long-lasting significance has been social and moral. Remembrances of events like Knox can serve important social and moral purposes. Remembering can link the past with the present, connecting where the community has been with what it is today. Most people in northeastern Pennsylvania have some personal tie to coal mining, making the Knox disaster a tragedy that so many could relate to. Remembering also takes pivotal events out of the past and places them in the present, where people can reconsider and re-interpret their causes and consequences. In so doing, remembering highlights the social forces and powerful people that have dominated our past and, for better or worse, have helped shape the present. In short, events like Knox bind us together as a community.

As we commemorate the forty-fifth anniversary of the Knox calamity in 2004, I recommend that we do four things. First, grieve with the victims' families and with others whose ancestors were among the multitudes who died underground or otherwise expired as a result of their toil in anthracite. Second, acknowledge our forebears who endured injury, insecurity, violence, and even death to provide a better life for us. Third, educate young and old alike about the rich yet often-sorrowful history of hard coal—it is our heritage and we should be proud of it. And fourth, act to transform the present by making certain that illegalities and corruptions (the fundamental causes of the Knox disaster) are not tolerated in any form in the anthracite region—a transformation that will provide a fitting and enduring legacy for the twelve men who perished at Knox forty-five years ago.

## Essay: "Why We Need to Understand and Appreciate Anthracite History"

### by Robert P. Wolensky

(Earlier versions of this essay appeared in the *Citizens' Voice,* and in the *Anthracite History Journal,* Summer 2003)

Leaving anthracite history aside for the moment, let me ask: why do we need to study any history? The answer to this question is much more complicated than the old saw that learning history prevents us from repeating the mistakes of the past. Certainly we don't want to repeat earlier blunders, but I would argue that historical understanding can help especially young and middle-aged people deal with two issues that seem to cause extensive social and psychological distress nowadays: finding meaning and purpose to life.

In my view, the contemporary age presents some major obstructions to meaningful and purposeful living. One problem is that too many of us do not know how to live in the present because, at virtually every moment, we are rushing toward the future. The pace of daily life has increased dramatically in recent decades.[5] We regularly check our watches because, "I have to be someplace." And then, when I get to this future point, the process is repeated because, "I have to dash to someplace else." We have somehow created a culture of anticipation, anxiety, and future orientation—a culture of "what's next?"

Fig. 77. Commemorating the forty-fifth anniversary of the Knox Mine Disaster on January 22, 2004. (Courtesy John Dziak)

I am reminded of British Cardinal John Henry Newman's assertion that death was not what he feared most; rather it was never having learned how to live. I believe he meant "how to live without anxiety, happily in the present."

Another contemporary malady is that we often seek meaning and purpose in material consumption. The good life has become the "goods" life. Many of us suffer from affluenza, a cultural "disease" resulting from our fervent consumerist predilections.[6] While things provide no more than a quick fix for what ails us, they actually reinforce anxieties about what to consume next or "am I consuming the right goods?" They also distract us from family obligations and civic participation, and from cultivating our inner well-being. Are we more than what we wear, drive, eat, drink, or live in? I believe and hope so.

If I am correct in saying that we have great difficulty living in the present and are speedily moving to the future, how can we possibly pay any attention to the past—to the realm of history? Yet, I believe it is precisely in a study of the past where we can find the wherewithal to cultivate a meaningful and purposeful life in the present.

If I am correct in saying that we are much more concerned with consumption then how can we overcome the problems associated with affluenza? Again, I believe we can alleviate this ailment by finding deeper meanings, inner strengths, and a more fulfilling purpose by seeking knowledge about who we are, where we came from, and how our ancestors and neighbors confronted the problems of their times.

And this brings me to anthracite. For too long, anthracite has been a history to forget. Thankfully, over the past ten to fifteen years more and more people from hard coal's cities and towns have come to realize the region's incredibly rich history. It is a story replete with examples of hard work, brave immigration, family solidarity, fearless unions, ethnic and religious solidarity (and intermarriage), economic advancement, community resilience, and educational attainment. It is our own unique version of the American chronicle.

Certainly there is the dark side too: accidents, strikes, deaths, losses, crimes, injustices, corruptions, environmental scars, alcohol and drug abuses, and calamities such as the Avondale and the Knox mine disasters. Yet these are also part of our history and I believe that we must accept these problems so as to build and learn from them. Our tribulations can and should become a source of personal and social strength. If our ancestors did what they did, triumphed over

Fig. 78. The Huber Breaker, Ashley, Pennsylvania, photo circa 1960, the last standing breaker in the Wyoming-Lackawanna anthracite coal field.

Fig. 79. The Huber Breaker, Ashley, Pennsylvania, today. Citizens are seeking to preserve the breaker as a historical and educational center. (Courtesy William Best)

their difficult circumstances, then we can too. It won't be easy, but our tasks can hardly be more difficult than theirs. And remember, they had us in mind. They wanted to make this community a more socially and economically just place not only for themselves but also for their descendents. We are the beneficiaries and we own a debt to those who have given us so much.

While we have made notable progress in studying and acknowledging our past, we still have a long way to go. Anthracite studies are rarely taught in local schools or colleges. Although industrial tourism has been developed to some extent in Scranton, Eckley, Lansford, and Ashland, much more still needs to be done. When can we tour a coal breaker, a garment shop, a shoe factory, a silk mill, a strip mine, and a cigar factory? We are still throwing away many of the papers of our houses of worship, ethnic organizations, civic associations, coal companies and other businesses. The papers of one of the region's major coal enterprises—the Glen Alden—are in dire need of preservation. The few remaining historical gems of the anthracite era, such as Concrete City, the Ashley Planes, the Huber Breaker (figures 78 and 79), and the St. Nicholas Breaker need to be saved before the wrecking ball does its deed.

My point is that if we are going to cultivate our personal and social strengths by cultivating our heritage, then we have to preserve and educate about that heritage. This, it seems to me, will provide real food for hungry souls looking for meaning and purpose in an often superficial, materialistic, and hurried society. And, in the process, we may even learn how to avoid some of the mistakes of the past!

## NOTES FOR CHAPTER SEVEN

1. DeGerolamo, who served as financial secretary of UMWA Local 9874, commented on his reading of the eulogy: "I thought I was going to break down before I finished; every word I issued became harder and harder with a tremble in my voice. By the time I finished, the tears were streaming down my cheeks. It was the most solemn day of my life. When I got through, a local priest said the Lord's Prayer." (Autobiography, 1980, p. 55)

2. See Donald Baloga's oral history account regarding the clock in chapter five.

3. The lyrics for Lex Romane's song were presented in chapter six.

4. The Knox Mine Disaster documentary debuted on WVIA-TV, public television in northeastern Pennsylvania, in 1989. See note 1, chapter five, for more information on the program.

5. Research by Juliet B. Schor documents the overworked and overspent predicament in contemporary American life. See her books *The Overworked American: The Unexpected Decline of Leisure* (New York: Basic Books, 1991); and *The Overspent American: Upscaling, Downshifting, and the New Consumer* (New York: Basic Books, 1998).

6. The term Affluenza comes from the title of a documentary produced by KCTS/Seattle and Oregon Public Broadcasting that aired on PBS in September 1997. For more information see www.pbs.org/kcts/affluenza See also the sequel *Escape from Affluenza*, which also debuted on PBS, as well as the work by the Milwaukee-based organization called The Affluenza Project (www.affluenza.com).

# Glossary[1]

**anticline.** A flexure or fold in which the rocks on the opposite sides of a fold dip away from each other, like two legs of the letter A. The effect can also resemble a saddle.

**Baloga slope.** An internal slope at the Ewen Colliery connecting the Pittston and the Marcy Veins.

**barrier pillar.** A solid block of coal left between two mines as security against accidents arising from an influx of water or other underground accident.

**breaker.** The tall surface building in which the coal is broken, sized, and cleaned for market.

**Big Vein.** Another name for the Pittston Vein, the top-most of seven veins in the Pittston area.

**cage.** The shaft elevator that moves men, coal, and materials into and out of a mine.

**contracting-leasing system.** Beginning in the 1920s, the large anthracite coal companies that controlled the mineral rights in the northern anthracite field began (sub)contracting sections of their mines to individual entrepreneurs who hired crews of up to twenty workers. The entrepreneurs often boosted production and lowered costs by taking coal in unsafe places, ignoring safety procedures, and violating the collectively bargained agreement with the United Mine Workers of America. During the 1930s the large firms—with the Pennsylvania Coal Company leading the way—went beyond contracting and began leasing entire collieries to independents that were now incorporated. The Knox Coal Company obtained its first lease from the Pennsylvania Coal Company in 1943. Some of the contractors and lessees had alleged ties with the area's organized crime syndicate, including the Knox Coal Company. For more details on this arrangement See Robert P. Wolensky, Kenneth C. Wolensky, and Nicole H. Wolensky, *The Knox Mine Disaster,* Harrisburg: PHMC, 1999.

**crosscut.** a passageway driven at right angles to a mined-out chamber to connect it with a parallel chamber for the circulation of air.

**dog hole.** A meager and often illegal (or "bootleg") mining operation run by a small crew of workers.

**droppers.** Droplets of water that fall from the roof of a mine chamber.

**Eagle Air Shaft.** Created by the long defunct Eagle Coal Company, it was the only opening available for the thirty-three mineworkers who escaped through it during the Knox Mine Disaster.

**excelsior.** A baled wood product resembling straw, used in the effort to plug the hole in the Susquehanna River caused by the Knox Mine Disaster.

**face.** The "front" of a workplace where the coal is mined.

**gangway.** the main underground road or passageway in an anthracite mine, it was laid with tracks for the movement of coal cars.

**gob.** 1. To store underground, as along one side of a working place, the rock and refuse encountered in mining; 2. The material so packed or stored underground.

**heading.** A gangway, entry, or airway.

**Hoyt Shaft.** An opening at the Ewen Colliery of the Pennsylvania Coal Company, it was used by the Knox Coal Company as part of its lease.

**jalopy.** Miners' slang for a type of chain conveyor used to move coal from the face to where it could be loaded.

**Joy Loader.** An automatic mechanical loader that scooped up the coal and loaded it into a coal car, manufactured by the Joy Company. Few actually in use in anthracite because of unique mining conditions. Knox Coal Company had one Joy Loader.

**loadeds.** Mineworkers' slang for cars filled with coal, they were taken to the cage first by a mule and later by a "motor" for the trip to the breaker where the coal was processed.

**laborer.** An uncertified mineworker who performs various tasks such as loading coal or rock, topping coal cars, carrying supplies etc.

**main road.** The most important haulage road in a mine, also known as the main heading road or gangway.

**main heading road.** A term referring to the gangway at the Ewen Colliery of the Pennsylvania Coal Company (and apparently at very few other operations).

**manway.** A narrow passage used by workers to travel between workings.

**Marcy Vein.** The second vein of coal in the Pittston area, directly below the Pittston or "Big" Vein.

**May Shaft.** The main opening at the Ewen Colliery of the Pennsylvania Coal Company, used by the Knox Coal Company as part of its lease.

**miner.** While the term has been used to apply to a broad category of workmen, it most accurately refers to a state-certified worker (one having "mining papers") who has the authority to plan work, set and fire a dynamite charge, and perform other exclusive tasks. Company miners worked steadily for the same firm performing various tasks, while contract miners headed crews of one to four laborers having the main task of mining coal.

**motor.** Generally applied to devices that take electrical power and transmits it as mechanical power to a machine. As used in anthracite, a haulage engine or locomotive operated by electricity or compressed air.

**motorman.** Driver of a mine motor.

**nipper.** The entry-level underground job for a boy, it involved sitting near a door and opening and closing it to allow for the passage of mine cars.

**pans.** A term applied to a shaker conveyor trough and less frequently to a chain conveyor trough, which were types of mining machines.

**Pennsylvania Coal Company.** Owner of the Ewen Colliery and many other collieries, the firm leased mineral rights to the Knox Coal Company beginning in 1943.

**pillars.** Blocks of coal within a mine resulting from a first mining using the "room-and-pillar" method. They were usually removed or "robbed" in a second mining, hence the term "robbing the pillars," typically an acceptable and legal undertaking.

**pitch.** The rise of a coal vein or grade of an incline.

**Pittston Vein.** The top-most vein in the Pittston area, also known as the "Big" Vein because of its thick and rich coal.

**place.** Common term for a work place or mine chamber.

**props.** A wooden or metal temporary support for the roof.

**Red Ash Vein.** The bottom-most vein of seven veins in the Pittston area.

**rib.** The sidewall in a gangway or mine chamber.

**robbing the pillars.** The removal of blocks of coal from a mine in a second mining. See pillars.

**rockmen.** Mineworkers who quarry entry tunnels from the outside into a mine, or within a mine from one vein to another.

**roof.** The rock "ceiling" of a mine chamber.

**room-and-pillar method of mining.** A method used in the northern field whereby blocks of coal were left between mined out chambers. The blocks, called pillars, were used to support the roof until first mining was finished whereupon a second mining was undertaken to remove the pillars.

**Schooley Shaft.** Located on the east side of the Susquehanna River in Exeter, it served as the main entrance and exit at the Schooley Coal Company, whose workings were later owned by the Pennsylvania Coal Company and then leased by the Knox Coal Company.

**shaker chute.** Metal troughs that operated mechanically in a shaking fashion, used for loading of coal into mine cars.

**shaft.** A vertical mine opening used 1. to transport men and mine cars into and out of a mine via the shaft elevator; 2. to discharge water through pipes; and 3. for the passage of air into a mine.

**shifting shanty.** A heated building on the colliery grounds used by mineworkers to change clothes and wash up or shower after a shift of work.

**slope.** An inclined mine roadway usually driven from the surface and laid with rails to access coal veins below. A rock slope is driven between strata within a mine to connect veins.

**spad.** A horseshoe nail with a hole in the head, or a similar device for driving into mine timers, or into a wooden plug fitted into the roof, to mark a survey station.

**topper.** A laborer in a mining crew whose job is to complete the filling of a coal car by piling chunks of coal on to the car until it reaches the acceptable height.

**workings.** Common term for underground mining operations.

### NOTES TO GLOSSARY

1. The glossary draws upon mining terms contained in the Hudson Coal Company's book, *The Story of Anthracite,* published by the company in 1932; and in *A Dictionary of Mining, Mineral, and Related Terms,* compiled and edited by Paul W. Thrush and the staff at the U.S. Bureau of Mines, Department of Interior, USGPO, 1968. Bill Hastie and Joe Keating also assisted with the glossary.

# Index

Aiello, Theresa, 109, 146
Alaimo, Dominick, 13, 15
Albert Aston, 13
Alteri, Mary, 148
Altieri, Frank Samuel, 148
Altieri, Samuel, 7, 85, 95, 108, 147–53
Altieri, Vincent, 109, 148
Amico, Matthew, 240
anthracite coal region, map, 2
Anthracite Heritage Museum, 234, 249
*Anthracite History Journal,* 226, 248, 251
Argo, Anthony, 13
Arone, Anthony, 90–92
Ashley Planes, 254
Avondale Mine Disaster, 240, 246, 253
Baldricia, Rena, 207–10
Baloga, Caroline, 101, 109, 128–37, 170–74, 174–82, 202–3
Baloga, Donald, 173, 174–82, 202–3, 245, 255
Baloga, John, 7, 95, 170–74, 174–82, 202–3
Bartoli, Art, 238
Bendick, Fr. John J., 249
Bernardi Coal Company, 80
Best, William, 194, 227, 236–42, 244–45
Bianconi, Adrian Mark, 224
Blara, John, 95

Bohn, Fred, 3–4, 25–26
Boroski, Edward, 84–87, 95, 98
Boyar, Benjamin, 7, 94–95, 153–59
Boyar, Richard 153-59
Burke, Emmet, 212
Burns, Francis Jr., 159–64
Burns, Francis Sr., 7, 67, 74, 84, 94–95, 157, 159–64
Burns, Thomas, 34–38, 43
Calvey, Audrey, 170–74, 202–3
Cappellini, Judge Gifford, 234
Cappellini, Rinaldo, 16
Cawley, Paul, 95
Cecconi, Fred, 31–32, 35, 38, 40, 42
Chickson, Rose, 196
Ciampi, Mrs. Herman, 148
Cigarski, Stephen, 95
Concrete City, 242, 254
Connelly, Daniel, 43
Cotter, Michael, 232, 236
Creasing, Stuart Co., 3, 7, 139, 182
Curran, Msgr. John J., 16
Czechoslovakia, 170
Davenport, Linda, 145, 153–59
DeAlba, Sam, 152–53, 229–36
DeGerolamo, George, 211, 227–29, 255
Domoracki, Frank, 3–4, 22, 24, 26
Donegal Weavers, 212–14, 226
Dougherty, John, 212
Dougherty, Robert, 13, 15, 130
Dunn, Chester, 29–33
Dziak, John, 22, 194, 237, 239, 242–48, 252,
Eagle Air Shaft, 51–54, 57, 64–65, 69–71, 74, 77–79, 82, 84–85, 87–88, 90–92, 98, 102, 172, 214,

238, 241, 246
Economy Stores, 187
Elko, John, 58, 59
Fabrizio, Louis, 13, 15, 130, 141, 149–51, 175–76
Featherman, Charles, 4, 7, 95, 116–19, 136
Featherman, Ervin, 136, 137
Featherman, Opal, 116–19
Featherman, Sherry, 117
Ferrare, Ann 108, 147–53
Ferrare, Frank, 147–53
Finn, Ed, 194
Flood, Rep. Daniel J., 114
Foglia, Mrs. Nicholas, 148
Forty Fort, 158
Francik, Joseph, 68–71, 95, 98
Fries, Ralph, 15
Funke, Erika, 194, 249
Gadomski, John, 63, 71, 75, 75–80, 82, 95, 98
Gadomski, Mrs. John, 79
Gizenski, Alfred, 137–43
Gizenski, Ida, 117, 137–43
Gizenski, Joseph "Tiny," Sr. 4, 7, 26, 95, 117, 185
Gizenski, Joseph Jr., 137–43
Glen Alden Coal Company, 254
Goodford, Edward A., 204
Greater Pittston Historical Society, 194, 234, 242–43, 249
Griglock, Irv, 238
Groves, Robert, 4, 13, 15, 28–31, 33, 38, 40, 58, 82, 116, 114–15
Gustitus, John, 75, 76, 82, 88, 95, 98
Hague, William, 90–92
Handley, Frank, 32–35, 40–41, 43, 46, 48, 54, 124
Hanusa, Michael, 48
Harvan, George, 8, 102
Hastie, William, 1, 7, 10, 21, 57–58, 60, 94, 194, 239, 240, 248
Holton, Lillian, 186, 191
Hopkins, Joseph, 92
Huber Breaker, 227, 236, 238, 241, 253–54
Hudson Coal Company, 15–16
International Ladies Garment Workers' Union, 135, 152, 202
Irish Balladeers, 204–6
Italy, 234
Jamieson, James, 38, 40, 43, 92
Jeffries, James, 92
Jewish Community Center, 152
Jones, Dolores Tomaszewski, 110
Jones, Joseph, 212
Kachinski, Joseph, 95
Kaloge, Joseph, xi, 45–46
Kanjorski, Rep. Paul, 234
Kaveliski, Henry J. 127
Kaveliskie, Dominick, 7, 95, 105–6, 127
Kaveliskie, Mrs. Dominick, 105–6
Keating, Joe, 246
Kehoe, John Sr., 128, 143
Kehoe-Berge Coal Company, 127
King's College, 159, 161, 164,
Kluger, Alan, 115
Knox Coal Company, 10–13, 35, 43, 46, 51, 58, 60, 66–67, 81, 115, 130, 150, 155, 163, 172–73, 175, 180, 193, 248
Knox Mine Disaster
  anniversaries, 194, 233–34, 241,

245, 248–52
causes, 11–17
cultural meaning, 248–55
documentary film, 194, 249, 255
economic impact, 17
further reading, 20
investigations, 11–17, 96–97
markers, 194, 229–36, 236–42, 249
Memorial Committee, 229–36
memorial service, 173, 189, 232, 235, 249
model, 246–47
monument, 152–53, 173, 223, 229–36
musical compositions, 200–1, 204–6, 212–14, 214–15, 220–23, 23–25
pumping afterwards, 145
plugging the hole, 10
reform legislation following, 17–18
social and cultural implications, 18, Epilogue
victims, 7
Kopcza, Joseph, 39–43
Krywicki, Anthony, 95
Kulesa, Chester, 219
Lackawanna Coal Mine Tour, 238
LaFratte, James, 52–53, 58, 95
Langan, Fr. Edmund, 5, 232
Lehigh Valley Coal Company, 10, 16
Lippi, August J., 13–15
Lombardo, Mayor Michael, 234
Lucas, Michael, 26–29, 38, 48, 116, 123–26, 168
Ludzia, Frank, 95
Lukasik, Stephen, 9, 12, 89, 96, 99, 104, 213
Luzerne County Community College, 249
Lyons, Clementine, 202
Mackachinus, Michael, 95
Maloney, Thomas, 16
Mangan, Jean, 232, 236
Marino, John, 230, 236, 238–39
Marscio, Louis, 95
Matheson, Min, 135, 143–44, 202
Mazur, George, 71–75, 76, 95, 98
McGroarty, Fr. Hugh, 194, 240
McKeever, Edgar, 216
McKeever, Erik, 195, 215–20
Michulis, Charles, 95
Moore, Joseph, 95
Murman, Frank, 195, 200–1
Murman, Tom, 195, 200–1
Murtha, Nora, 232, 236
Musto, Rep. James, 25
Musto, Rep. Raphael, 232, 234
Myers, George H., 206
Nelson, P. J., 198
Northeastern Pennsylvania Oral and Life History Project, x
Obsitos, Michael, 46–48
Ogin, Anita, 145, 182–92
oral history, x–xi
organized crime, 12
Orlando, Pep, 230–31, 236
Orlowski, Francis Robert, 110
Orlowski, Frank, 1, 3, 7, 95, 104, 218–20, 226
Orlowski, Mrs. Frank, 106–8
Orlowski, Theresa, 104, 107, 112
Ostrowsi, Donna, 182–92
Ostrowski, Eugene Jr., 182–92

Ostrowski, Eugene Sr., 3–4, 7, 95, 182–92
Ostrowski, Theodosia, 184–86, 189–90
Paluske, Joe, 3
Pancotti, Amadeo "Paul," 4, 5, 54, 57–60, 62, 76, 78–80, 85–86, 95, 208, 223, 226
Parente, Yolanda, 109, 147–53
Payne Colliery, 80
Pennsylvania Coal Company, 10–12, 15–16, 33, 43, 49, 51, 58, 105, 130, 155, 157, 160, 163, 175, 226
Pennsylvania Department of Mines and Mineral Industries, 9, 13
Pennsylvania General Assembly, 13
Pennsylvania Historical and Museum Commission, 232, 236, 249
Pernot, Ray, 194
Piasecki, Charles, 13, 15
Pientka, John, 67, 95
Pittston Hospital, 84, 87–94, 112, 148–49, 163, 172, 177, 192, 242–44
Plymouth Historical Society, 240
poetry, Chapter Six
Poland, 158
Proshunis, Charles, 95
Ramage, Merle, 67, 95, 124
Randazza, Louis, 58–59, 95
Ray Stephens, 212, 214
Receski, William, 12–13, 15, 92, 126
Reese, Mrs., 198
Remus, Anthony, 43–45
Renner, Fritz, 15
Retondaro, Angelo, 95
Richard Boyar, 153–59

Roberts, Ellis, 45, 98
Robshaw, Robert, 40, 43
Rogers, Charles, 204
Roman, Stanley, 75, 76, 80–84, 95, 98, 249
Romane, Lex, 220, 221-23, 255
Salvation Army, 110, 146
Saporito, Martin, 95
Sciandra, John, 13
Sciandra, Josephine, 13, 15
Shane, George, 31, 40, 95
Shane, Joseph, 31
Sinclair, Frances O., 101, 110–12
Sinclair, William, 7, 26, 28, 37–38, 123–24, 126
Slempa, Peter, 48
Smelster, Albert, 95
Solarczyk, Joseph, 95
Soltis, Joseph, 58–59, 78, 90, 95
Sporher, George, 115
St. Joseph's Catholic Church, 123, 152–53, 173, 189–90, 223, 229, 232, 236–37, 249
St. Mary's Lithuanian Church, 186
Staliha, Charles, 178
Stark, Lea, 112–16
Statkiewicz, Leonard, 14
Stefanides, Christine, 120, 121
Stefanides, Daniel Jr., 121, 164–70
Stefanides, Daniel Sr., 7, 26, 28–29, 38, 46, 48, 95, 119–26, 164–70
Stefanides, Joseph, 115, 120, 123–26, 167–68
Stefanides, Michael, 166
Stefanides, Patricia, 120, 121, 123,
Stefanides, Richard, 166, 170
Stefanides, Stephanie, 119–23

Stefanides, Stephen, 169
Stefanovich, Marion, 199
Stella, Pacifico "Joe," 4, 21, 49–57, 64, 69, 74–80, 83, 85–87, 89, 95, 98, 126, 194, 226, 237, 240–41, 244–45, 249
Stephen Teller, 13
Stuccio, Jerome, 52–53, 58, 78, 95
Stupak, John, 42
Suchocki, Alfreda, 231, 233
Susquehanna River, 177–78
Swoyersville, 166, 168, 268
Taft-Hartley labor law, 13, 222
Talipan, Jean, 231, 233, 236
Thomas, Myron, 5, 49–52, 54, 59, 63–69, 74–79, 82–83, 85–86, 95, 98, 226
Thomas, Robert, 98
Tigue, Rep. Thomas, 233-34
Twin Shaft Disaster, 143, 246
United Anthracite Miners of Pennsylvania, 16–17
United Mine Workers of America, District 1, 12–13, 16, 187, 222
United Mine Workers of America, Local 8005, 13
United States Bureau of Mines, 13
Vella, Leo, 43
Waitcavage, Anthony, 37
Wascalis, Francis, 95
Watkins, Thomas, 38
Williams, Jack, 3, 4, 24–25
Williams, Violet, 92–94
Wolensky, Kenneth, 18, 232, 268
Wolensky, Nicole, 18, 268
Wolensky, Robert, 18, 248–55, 268
WVIA–TV, 194, 249, 255
Yeager, George, 212

## About the Authors

**Robert P. Wolensky** serves as professor of Sociology and co-director of the Center for the Small City at the University of Wisconsin-Stevens Point. A native of Swoyersville and Trucksville, Pennsylvania, he graduated from Central Catholic High School, Villanova University, and Penn State University. His current endeavor, working in cooperation with King's College in Wilkes-Barre, is to preserve and disseminate information about the more than six hundred interviews contained in the Northeastern Pennsylvania Oral and Life History Project.

**Nicole Wolensky** is a native of Stevens Point, Wisconsin. She received B.A. degrees in sociology and psychology from Marquette University in 2001 and an M.A. in sociology from the University of Iowa in 2003 where she is a Ph.D. candidate in sociology. Her research interests are in social psychology, family, labor history, sports, and religion. She lectures in social psychology at Iowa, and plans for a career in tertiary teaching. She is a member of the Iowa graduate student union (COGS), which is affiliated with the United Electrical Workers Union.

**Kenneth C. Wolensky** is a historian with the Pennsylvania Historical and Museum Commission. He specializes in industrial, labor, and public policy history. He has published and spoken widely on the history of the labor, industry, working class culture, politics, and public policy. He has an extensive background in public administration and is a faculty member for the American Studies program at Penn State University Capital College, a Distinguished Lecturer for the Organization of American Historians (OAH), and has served as a lecturer for the Pennsylvania Humanities Council.

Robert, Ken, and Nicole Wolensky co-authored *Fighting for the Union Label: The Women's Garment Industry and the ILGWU in Pennsylvania* (Pennsylvania State University Press, 2002), the first comprehensive study of the women's garment industry and its major labor union in the Keystone State. They also co-authored *The Knox Mine Disaster* (Pennsylvania Historical and Museum Commission, 1999).